A NEW LIMIT
MECHANICS

Dennis Theron Lewis

A NEW LIMIT MECHANICS

To order additional copies of this book, contact:
Xlibris Corporation
1-888-795-4274
www.Xlibris.com
Orders@Xlibris.com
45255

CONTENTS

To the Loving Memory of
Vera May Harwood
1911-1991

PREFACE

Around the year 1983, I had resolved Galileo's famous Pisa experiment. I wrote a paper about it, but the paper was rejected by a couple of scientific journals. After this important empirical resolution, I began to think that I might make a contribution to physics.

I then directed my attention to Albert Einstein's special theory of relativity. Around that time, 1983 to 1989, I began to use the concepts of "limited", and "limit", which were to be used instead of the outdated concepts of "relativity", and "absolute". Eventually, after much writing, thinking, and rejection of "ideas" after "ideas", I finally arrived at a small book, (the one you have in your hands)—and which purports, most sincerely, to replace Albert Einstein's special theory of relativity.

ABSTRACT

Albert Einstein's original special relativity theory is based primarily on his two postulates. And his two special postulates are based directly upon the original Newtonian first law of motion, which is a classical inertial law of motion. And then, in this book, I will propose and advance a new and revised Newtonian first law of motion, which is a more complete Newtonian first law of motion.

And then, I am rejecting Albert Einstein's two special postulates, and replacing them with two new and revised postulates; which are based directly upon this new and revised Newtonian first law of motion.

And in this sense, in this book, I will propose and advance a new non-relativistic mechanical theory (a new limit mechanics). And thus, I am rejecting, completely, all the mechanical results (all the dynamic and kinematic results): of Albert Einstein's original special relativity theory.

And in this same book, I will offer four new limited force measuring formulas, other then the one special relativity Newtonian measuring formula, offered by Albert Einstein in his special relativity theory.

And in this same book, I will offer four new limited momentum measuring formulas, other than the one special relativistic momentum formula, offered by Albert Einstein in his special relativity theory. And I will also offer four new lower-limit momentum formulas, and also offer four new lower-limit force measuring formulas.

INTRODUCTION

In the year 1905, Albert Einstein, a young European physicist, had proposed a new and remarkable physics theory, originally titled: *On the Electrodynamics of Moving Bodies*; and now entitled: The Special Theory of Relativity[1]. And according to his special relativity theory, Albert Einstein made a bold proposal to replace the Newtonian dynamics and kinematics, which involved absolute conceptions[2], such as absolute time, place, motion, and space—by his special relativistic conceptions.[3]

And in his 1905 paper, Albert Einstein advanced two special postulates[4] for his special relativity theory, which are:

Postulate (1)

The basic laws of physics are the same (invariant) for all inertial motive (non-accelerating) relative observers, or for all inertial relative reference frames.

Postulate (2)

In vacuum the speed of light, as measured by all inertial relative observers, or as measured by all inertial relative reference frames, is always a constant $+\underline{c}$, independent of the motion of the light source.

And by the use of his two special postulates, Albert Einstein derived the Lorentz-Einsteinian transformation equations. But which I am rejecting Albert Einstein use of the Lorentz-Einsteinian transformation equations as being empirically incorrect. And they are "empirically incomplete" because of their use by Albert Einstein, in his special relativity theory, will lead to

5

all of his special relativistic measuring formulas, which I totally reject, with respect to my new mechanical theory I will propose and advance in this book.

And I also reject Albert Einstein incomplete use (or handling of the Lorentz-Einsteinian transformation equations. Even though, the Maxwellian Electro-magnetic Equations are mathematically invariant under these same Lorentz-Einsteinian transformation equations: which is very important in Albert Einstein's special relativity theory.[5]

And Albert Einstein, in regards to his two special postulates, had used the special relativitic conceptions of: "inertial relative observers". And these special relativitic conceptions are empirically valid and are true conceptions, *in which the original Newtonian first law of motion holds for them*. And then, we see the fundamental basis of Albert Einstein's special relativity theory is Isaac Newton's original first law of motion,[6] which is a classical inertial law of motion.

We now have:

The Original Newtonian First Law of Motion

Every cosmic body, or every cosmic mass persevers, (continues and remains), in its state of inertial rest, in free vacuum space and time. Or has an inertial velocity, or inertial speed, displacement—in free vacuum space and time: unless it is compelled to change these inertial states by forces imposed.

And because of the original Newtonian first law of motion is the reason why Albert Einstein, and the many proponents of his special relativity theory, have freely discussed and freely considered: *all kinds of inertial states are freely given by Nature*. And these relativistic conceptions are based directly upon their understanding and acceptance of the original Newtonian first law of motion.

But in this book, I am rejecting the original Newtonian first law of motion as being an incomplete and inconsistent (classical) inertial law of motion. And instead of the original Newtonian first law of motion, I will offer a new and revised and complete and consistent Newtonian first law of motion. And which is a more empirically and complete inertial law of motion in comparison to the original Newtonian first law of motion.

I am also rejecting Albert Einstein's two special postulates (1), and (2), in regards to Albert Einstein's incomplete use of the two special relativistic conceptions of: "inertial relative reference frames", and of: "inertial relative observers"—with respect to my new mechanical theory that I am proposing and advancing in this book.

We also know that these two special relativistic conceptions are empirically valid and are true conceptions, according to which the original Newtonian first law of motion holds for them. And now, the importance of the original Newtonian first law of motion holds for Albert Einstein's special relativity theory is now recognized.

And by the use of his two special postulates (1) and (2), and including his incomplete use of the Lorentz-Einsteinian transformation equations. As well as by his incomplete use of the two special relativistic conceptions of: "inertial relative reference frames", and of: "inertial relative observers". Albert Einstein then derived and formulated[7] the many special relativistic measuring formulas. And these special relativistic measuring formulas are:

The special relativistic mass increase measuring formula:

$$+\underline{m} = (+\underline{m}_o / ((+1-(\underline{u}/\underline{c})^2)^{+\frac{1}{2}}, \text{ where: } (+\underline{m} \rightarrow +\infty).$$

(where the symbol $+\infty$ is the mathematical symbol for positive infinity). The special relativistic momentum measuring formula:

$$+P = +\underline{m}_o v / ((+1-(\underline{u}/\underline{c})^2)^{+\frac{1}{2}}$$

The special relativistic Newtonian measuring formula:

$$+F = (+\underline{m}_o \dot{v} / ((+1-(\dot{\underline{u}}/\underline{c})^2)^{+\frac{1}{2}}$$

The special relativistic time dilation measuring formula:

$$+\underline{t} = (+\underline{t}_o) / ((+1-(\underline{u}/\underline{c})^2)^{+\frac{1}{2}}, \text{ where: } (+\underline{t} \rightarrow 0).$$

The special relativistic kinematic length contraction measuring formula:

$$+\underline{L} = (+\underline{L}_o)(+1-(\underline{u}/\underline{c})^2)^{+\frac{1}{2}}$$

The special relativistic rest energy measuring formula:

$$\pm\underline{E} = (+\underline{m}_o\underline{c}^2) \text{ and: } \underline{E} = (+\underline{mc}^2)$$

The special relativistic total kinetic energy measuring formula:

$$+\underline{T}_c = (+\underline{m}_o\underline{c}^2)/(+1-(\underline{u}/\underline{c})^2)^{+\frac{1}{2}}-1)$$

And I am rejecting all of these many kinds of special relativistic measuring formulas, which were derived and formulated from Albert Einstein's special relativity theory, as being false measuring formulas, according to this new mechanical theory I am proposing and advancing.

And therefore, in this book, I have rejected Albert Einstein's original special relativity theory, according to my rejection of the original Newtonian first law of motion: which is the fundamental empirical basis for Albert Einstein's special relativity theory. And I have also rejected Albert Einstein's special relativity theory because of my complete rejection of his special relativistic measuring formulas. And in this book, I will offer a new non-relativistic mechanical theory, and thus, I reject completely, Albert Einstein's special relativity theory.

We now have a new and revised Newtonian first law of motion:

The New Revised Newtonian First Law of Motion (4)

Before every limited inertial cosmic mass, persevers (continues and remains), at a limited inertial cosmic rest mass state, in free vacuum space and time.

Hence, *a limited force had formerly been used upon a limited inertial velocity displaced mass +m*—in the opposite direction of the mass +m—to have set the mass +m *into a limited inertial rest mass state.*

Before every limited inertial cosmic mass, persevers (continues and remains), at a limited and constant right line velocity displacement, in free vacuum space and time.

Hence, *a limited force had formerly been used upon a limited inertial rest mass* $+\underline{m}_0$ (*and the instantaneous discontinuance of this same force*)—to have set the limited inertial rest mass $+\underline{m}_0$ *into a constant right line inertial velocity displacement.*

And whereas, all other external net cosmic forces upon these same limited inertial cosmic masses cancel out to zero effects.

And according to this new revised Newtonian first law of motion, I have used the two new empirical conceptions of: "limited forces", and of the "instantaneous discontinuances" of these same "limited forces". And also in regards to these same two new empirical conceptions, my new mechanical theory is then empirically valid and holds to this new revised Newtonian first law of motion.

This is because the fundamental empirical basis of my new mechanical theory is this revised Newtonian first law of motion, which is a more complete and consistent inertial law of motion. In direct comparison to the original Newtonian first law of motion. And the main reason why it is a more complete and consistent inertial law of motion, is because it makes use of the two new empirical conceptions: of "limited force", and of the "instantaneous discontinuance" of these same "limited forces". And which these two empirical conceptions fully and correctly describes, or explains, all kinds of inertial states in Nature. And my new mechanical theory will fully make use of these two new empirical conceptions.

And the fundamental differences between the original Newtonian first law of motion, which is the fundamental empirical basis of Albert Einstein's special relativity theory—is that I ask for the causes of all inertial states in Nature. And thus, in this sense, my new revised Newtonian first law of motion must then involve the two new empirical conceptions of "limited forces", and of the "instantaneous discontinuances" of these same "limited forces". And while, the original Newtonian first law of motion accepts all kinds of inertial states as freely given, or for granted, without inquiring into the "empirical reasons", or "causes", of all these inertial states in Nature.

And Albert Einstein, and the many proponents of his special relativity theory, freely discussed and freely considered; as well as freely accepting all of these many kinds of inertial states as given, or for granted. Without in need of further empirical explanations of the "empirical reasons", or "causes", for these same inertial states in Nature.

And no motion is freely given by Nature, whether it be inertial velocity motion, or non-inertial accelerating motion, without there being causes and forces for these two kinds of motion.

And these "causes", or "forces", does not only result in motive displacements for a real displaced mass +$\underline{m}$, but also results in "intense dynamic and kinematic effects"—that a real displaced mass +$\underline{m}$ will experience. (As caused by a limited force). And also would result in "intense thermo dynamic and kinematic effects", that this same real displaced mass +$\underline{m}$ will experience. (As caused by limited forces). And in this book, I will give formulas measuring these "intense dynamic and kinematic effects" that a real limited displaced mass +$\underline{m}$ will experience. And also in this same book, I will examine and conjecture about these "intense thermo dynamic and kinematic effects" that we can expect this same real displaced mass +$\underline{m}$ will also experience.

Postulate (5)

(a) All limited $\dot{v}$ (non-inertial) accelerating displaced cosmic masses—(as caused by the instantaneous dynamic actions of a (classical) limited force: +F_v)—instantaneously, within a time interval $\underline{t}_b$, when from (0 to 1) of one second—or of a larger time interval—are invariant and are equivalent with each other, with respect to all the known basic laws of physics.

(b) (All limited v (inertial) velocity displaced cosmic masses)—(as due, or as caused by the instantaneous discontinuance of the dynamic actions of a (classical) limited force: +F_v). Instantaneously, at the end of a time interval $\underline{t}_b$, when +$\underline{t}_b$ = 1 second, or of a larger time interval—are invariant and are equivalent with each other, with respect to all the known basic laws of physics.

Postulate (6)

In free vacuum space and time, the uniform scalar inertial speed limit +$\underline{c}$ of light, (independent of the motion of the light source), as measured by all limited observers, and with respect to a basic time interval of one second: is always a fundamental measured constant speed limit +$\underline{c}$ of Nature. And which can serve as a fundamental upper speed limit, or a lower speed limit, for all limited (inertial and non-inertial) motive displaced cosmic masses.

We see that the main differences between Albert Einstein's two special postulates ($\underline{1}$) and ($\underline{2}$), and my two new postulates ($\underline{5}$) and ($\underline{6}$)—is that I have rejected Albert Einstein's incomplete use of the special relativistic conceptions of "inertial relative reference frames", and of: "inertial relative observers". And which these same special relativistic conceptions were incompletely used in Albert Einstein's two special postulates ($\underline{1}$) and ($\underline{2}$). (See conclusion and appendices more about this matter).

And my new postulates ($\underline{5}$, $\underline{a}$, and $\underline{b}$, and $\underline{6}$), are more general, and more complete, in an empirical sense, than Albert Einstein's postulate ($\underline{1}$). This is because I have used the two new empirical conceptions of: "limited +$\dot{v}$ (non-inertial) accelerating displaced cosmic masses". As caused by the instantaneous dynamic actions of a (classical) limited force: F_v, instantaneously, within a time interval $\underline{t}_b$, from (0 to 1) of one second, or at the end of a larger time interval.

And because I used the new empirical conception of: "limited v (inertial) velocity displaced cosmic masses". As are due, or caused by the end of the instantaneous discontinuance of the dynamic actions of a (classical) limited force: F_v—instantaneously, at the end of a time interval $\underline{t}_b$, when $\underline{t}_b = 1$ second, or at the end of a larger time interval. That is, I treat both non-inertial, and inertial states, in my postulate ($\underline{5}$, $\underline{a}$, and $\underline{b}$)

And then, because of my rejection of Albert Einstein incomplete use of the two special relativistic conceptions of: "inertial relative reference frames", and of: "inertial relative observers". And in this case, my two postulates ($\underline{5}$) and ($\underline{6}$), are in actuality, an indirect revision of Albert

Einstein's two special postulates (*1*) and (*2*), but which does without the use of these same two special relativistic conceptions.

I will now derive a new vector velocity limit term +c which is needed for my new mechanical theory. And which this term: +c is a fundamental "derived constant limit", of Nature. And whereas, in direct comparison to the fundamental measured constant speed limit +c of light, in free vacuum space and time. We are now given a unit vector term: +1. And then:

$$\text{Lim}_{v \to c} + \underline{mv} = (1 \cdot c) = (+\underline{mc} = 0) \tag{7}$$

And where $|+c| = +\underline{c}$. And where the term: +1 is a unit vector term having uniform direction, and having a uniform magnitude of a scalar numerical +$\underline{1}$. And according to the limit notation (7), the vector dot product of the unit vector term: +1 into the uniform scalar constant inertial speed limit +$\underline{c}$ of light in vacuum—is then the new vector constant inertial velocity limit term: +c. And as both terms: +$\underline{c}$ (the scalar constant inertial speed limit of light in vacuum). And the term: +c (the vector constant inertial velocity limit), are both to be measured by the time unit of one second.

And this new vector constant inertial velocity limit term: +c has both uniform vector direction and uniform magnitude. And while the scalar limit speed term: +$\underline{c}$ has uniform scalar magnitude. And this is the case provided if and only if the constant inertial speed limit +$\underline{c}$ of light: remains a fundamental measured constant inertial speed limit of Nature.

And this vector inertial velocity limit term: +c is to be used with all "idealized cosmic masses", which have "fictitiously obtained", in their inertial velocity displacements, this vector inertial velocity limit +c, as it is measured by one second.

And the empirical reason why I have used the phrase: "idealized cosmic masses", is that no real displaced cosmic mass +$\underline{m}$ can never obtain this vector limit +c, as it is measured by one second. And this fact is based upon the "intense dynamic and kinematic effects", and also upon the "intense thermo dynamic and kinematic effects" that this same mass +$\underline{m}$ would experience—by the extreme actions of a incredible (classical) force: $+F_v$, impacting upon the same mass $+\underline{m}$.

And these "intense effects" are caused by the former dynamic actions of a (classical) limited force: $+F_v$—instantaneously, within a time interval of one second, from (0 to 1), of one second, or at the end of a larger time interval. And these "intense effects" are also due to the instantaneous discontinuance of this same force: $+F_v$, of the time interval of one second, or perhaps of a larger time interval.

And these "intense effects" are the empirical reasons why I have set: $(+\underline{mc} = 0)$, and $(+\underline{mc} = 0)$, where the term: $+c$ is the vector constant velocity limit of an idealized mass. And where the term: $+\underline{c}$ is the scalar constant speed limit of light in free vacuum.

And I will, in this book, give new dynamic force measuring formulas, and give new kinematic momentum measuring formulas. And which these "formulas" will measure the "intense dynamic and kinematic effects" that a real displaced cosmic mass $+\underline{m}$ will experience. And I will also, in this book, further discuss, examine, and conjecture about the "intense thermo dynamic and kinematic effects" that this same mass $+\underline{m}$ will experience. But I am unable to offer any formulas, or equations describing, or measuring these "intense thermo dynamic and kinematic effects" that this mass $+\underline{m}$ will experience.

And perhaps, this vector constant velocity limit $+c$ can take its own natural place with all the other constants of Nature. Such as, in comparison to the measured constant speed limit $+\underline{c}$ of light, in free vacuum. The vector constant limit term: $+c$ is a derived constant of Nature, and instead of it being a measured constant of Nature, it is a derived constant of Nature.

I am now giving a diagram (8) to shown the empirical relations between a limited (classical) inertial rest mass term: $+m$, and of a limited (classical) uniform Newtonian force formula term: $+m\dot{v}$—and of a limited (classical) momentum formula term: $+mv$. And the empirical basis for these terms are due to the instantaneous dynamic actions of a "limited force", and of the instantaneous discontinuance of this same "limited force".

We now have a diagram:

Diagram (8)

$$(+F_v \rightarrow +\underline{m}_o \rightarrow +\underline{m}\dot{v}) \rightarrow +\underline{m}v))$$

On the left hand side of this diagram (8), means that a (classical) limited force: $+F_v$ is "instantaneously impacting" upon a limited inertial rest mass $+\underline{m}_o$ instantaneously, within a time interval $\underline{t}_b$, from (0 to 1) of one second; or of a larger time interval. And thereby, the former limited rest mass $+\underline{m}_o$ is now experiencing an instantaneous limited acceleration displacement, or a continuous displacement, in free vacuum space and time. Or the rest mass $+\underline{m}_o$ is experiencing an instantaneous, or continuous, *limited force displacement*. And these events are happening as caused by the incredible dynamic actions of a (classical) limited force: $+F_v$, instantaneously, or continuously, within a time interval $\underline{t}_b$, from (0 to 1), of one second, or of a larger time interval.

And on the right hand side of this diagram (8), means that the former limited accelerated displaced mass $+\underline{m}$, has now, "instantaneously inertially slides" into being an instantaneous, or continuously, a limited inertial velocity displaced mass $+\underline{m}$, in free vacuum space and time. Or is experiencing an instantaneous, or continuous, limited inertial momentum displacement. And these events are happening at the end of the instantaneous discontinuance of the same (classical) limited force: $+F_v$—instantaneously, at the end of one second.

And by this diagram (8), I have shown the dynamic and the kinematic relations between a uniform (classical) limited inertial rest mass term: $+\underline{m}_o$. And also of a uniform (classical) limited non-inertial force formula: $+F_v$. And also of a uniform (classical) limited inertial momentum formula: $+\underline{m}v$. And thus, these dynamic and kinematic relations of this diagram (8) are based directly upon the two new empirical conceptions of: "limited forces"—and of the instantaneous discontinuances of these same "limited forces".

I will be using in this book, an inequality:

$$(0 < k < 3) \tag{9}$$

Having the unit of measurement of: 10^4 kilometers per one second—which is the speed limit of light, in free vacuum.

In this book, when I use the conception of: "limited" what I mean by it is that for all limited motive displacements, and for all limited motive cosmic masses: which do not equal, nor obtain all limit motive displaced cosmic masses.

And these "limited motive cosmic masses" have the limited motive displacements of: +v, uniform inertial *limited vector velocity displacements*. And also of: $+\dot{v}$, a uniform non-inertial *limited vector acceleration displacements*.

And when I use the concept of: "limit", and which I mean by it, is that for all limit motive displacements, for all limit cosmic motive masses, which are the natural "lower-limit" for all limited motive displaced cosmic masses.

And also for: $(+\dot{c} = 0)$, and $(+\dot{c} = 0)$, are lower-limits displacements. We must also note that both of these two terms are of a zero second. But the terms: +c, and +c, are to be measured by a time interval of one second.

And it is in this sense the reason why I have set the limit lower-force formula: $+F_c$, where: $+F_c = (\underline{m}\dot{c} = 0)$; and the lower-force formula: $+\underline{F}_c$, where $+\underline{F}_c = (+\underline{m}\dot{c} = 0)$.

And where the zero value "0" in these two lower-force formulas indicates that this same inertial mass $+\underline{m}$ is non-existent in a mathematical sense, and in a physical sense. In a mathematical sense the first derivative of a constant number is a zero "0". And in a physical sense, the incredible and intense dynamic actions of a limited (classical) force upon the same mass $+\underline{m}$ would result in extreme thermo dynamic and kinematic effects as caused by this limited (classical) force: $+F_v$, and also by the limited force: $+\underline{F}_v$. We should note that it is accepted by physicists that no mass, whether earth bound mass, or cosmic mass, can never obtain the speed limit $+\underline{c}$ of light, in free vacuum.

Also, according to the limit lower-momentum formulas: $+P_c$, where $+P_c = (+\underline{m}c = 0)$. The intense and incredible dynamic and kinematic actions of a limited (classical) force: $+F_v$, would result in the same mass $+\underline{m}$ of these two formulas: into "intense implosions/intense explosions" of the same mass $+\underline{m}$. It is considered that no matter how extreme a force is, the mass would be converted into hot burning fragments of mass and energy being dispersed throughout the cosmos.

And these "extreme intense effects" are happening to this same mass $+\underline{m}$, instantaneously, within a time interval $\underline{t}_b$, from (0 to 1), of one second, or continuously, within a larger time interval. And we also should note that the limit lower-force formulas: $(+F_c = 0)$, and: $(\underline{F}_c = 0)$, serves as a lower-force limit for all limit force formulas. And that the lower-momentum formulas: $(+P_c = 0)$ and: $(+\underline{P}_c = 0)$, serves as a lower-momentum formulas for all limit momentum formulas.

In concluding this introduction, I should note that in the next following two sections of this book, I will offer four new limited dynamic $+F_v$ force measuring formulas. And I will offer four new limit dynamic $+F_c$ lower-force measuring formulas. That is, other than the one special relativistic Newtonian force measuring formula: $+F = +\underline{m}\dot{v} = (+\underline{m}_o\dot{v} / /(+1-(\underline{u}/\underline{c})^2)^{+\frac{1}{2}}$, offered by Albert Einstein in his special relativity theory.

And I will also offer four new limited kinematic $+P_v$ momentum measuring formulas. And I will also offer four new limit kinematic $+P_c$ lower-momentum measuring formulas. That is, other than the one special relativistic momentum measuring formula: $+P = +\underline{m}v = (+\underline{m}_o v / (+1-(\underline{u}/\underline{c})^2)^{+\frac{1}{2}}$.

THE NEW DYNAMIC FORCE MEASURING FORMULAS

In this section I will formulate the four new limited dynamic $+F_v$ force measuring formulas by using the limited (classical) Newtonian measuring formula: $+\underline{m}\dot{v}$. And also by using the formula: $+\underline{m} = (+\underline{V})(+\underline{d})$, where the term : $+\underline{V}$ is volume of the mass $+\underline{m}$, and where the term: $+\underline{d}$ is density of the mass $+\underline{m}$. I will also be using a limiting non-inertial factor: $(+1-(\dot{\underline{u}}/\underline{c})^2)^{+\frac{1}{2}}$. But I have not attempted to derive this limiting non-inertial factor from the two postulates I have offered in the new postulates (5) and (6). And in this book it must be accepted as it is.

And I will also formulate in this section the four new limit dynamic $+F_c$ lower-force measuring formulas by using a new limit (classical) lower-force measuring formula: $(+\underline{m}\dot{c}=0)$, where the term: $+c$ is the constant inertial *vector velocity limit term*. And I will also make use of the formula: $+\underline{m} = (+\underline{V})(+\underline{d})$. And I will also use the formula: $(+\underline{m}c = 0)$, where this formula is the lower-momentum scalar and limit inertial and constant formula. And where the term: $+\underline{c}$ is the scalar constant inertial speed limit of light in vacuum. And I will make use of the limiting non-inertial factor: $((+1-(\dot{\underline{c}}/\underline{c})^2)^{+\frac{1}{2}} > 0)$, as this limiting non-inertial factor is derived from:

$$\text{Lim}(+1-(\dot{\underline{u}}/\underline{c})^2)^{+\frac{1}{2}} = ((+1-(\dot{\underline{c}}/\underline{c})^2)^{+\frac{1}{2}} > 0) \quad \dot{\underline{u}} \to (+\dot{\underline{c}} = 0) \tag{10}$$

We now have, (in regards to a mass term: $+\underline{m}$), the two new limited force measuring formulas:

$$(+\underline{m}\dot{v})/(+1-(\dot{\underline{u}}/\underline{c})^2)^{+\frac{1}{2}} = +F_v' \tag{11}$$

And:

$$(+\underline{m}\dot{v})(+1-(\dot{\underline{u}}/\underline{c})^2)^{+\frac{1}{2}} = +'F_v \qquad (12)$$

And where $|+\dot{v}| = +\underline{\dot{v}}$. And where the new limited dynamic $+F_v$ force measuring formula (11) measures the instantaneous *acceleration-force displacements* (of increase/of decrease for a limited non-inertial accelerating displaced mass $+\underline{m}$. And which this same mass $+\underline{m}$ is being "instantaneously impacted", or "continuously impacting" by the dynamic actions of a (classical) limited force: $+F_v$. And these "dynamic actions" are happening, instantaneously, within a time interval $\underline{t}_b$, form (0 to 1) of one second, Or are happening continuously, within a larger time interval.

And where the new limited dynamic $+F_v$ force measuring formula (12) measures the instantaneous *mass-force displacement* (of increase/ of decrease): for a limited non-inertial accelerating displaced mass $+\underline{m}$. And which this same mass $+\underline{m}$ is being "instantaneous impacted", or "continuously impacting" by the dynamic actions of a (classical) limited force: $+F_v$. And these "dynamic actions" are happening, instantaneously, within a time interval from (0 to 1), of one second. Or are happening continuously, within a larger time interval.

We now have, (in regards to a volume term: $+\underline{V}$, and a density term: $+\underline{d}$, with respect to a mass term $+\underline{m}$., with respect to the formula: $(+\underline{V})(+\underline{d}) = +\underline{m}$, a scalar mass term). We then have the other two new limited dynamic force measuring formulas:

$$(+\underline{V}\dot{v})/(+1-(\dot{\underline{u}}/\underline{c})^2)^{+\frac{1}{2}} = +F_v^* \qquad (13)$$

and:

$$(+\underline{d}\dot{v})(+1-(\dot{\underline{u}}/\underline{c})^2)^{+\frac{1}{2}} = +^*F_v \qquad (14)$$

And where $|+\dot{v}| = +\underline{\dot{u}}$. And where new limited dynamic $+F_v^*$ force measuring formula (13) measures the *instantaneous volume-force displacement* (of decrease/ of increase): for a limited non-inertial accelerating displaced mass $+\underline{m}$. And which this same mass $+\underline{m}$ is being "instantaneously impacted", or "continuously impacting" by the dynamic actions of a (classical) limited force: $+F_v$. And these "dynamic actions" are happening, instantaneously, within a time interval $\underline{t}_b$, from (0 to 1), of one second. Or happening continuously, within a larger time interval.

And where the new limited dynamic $+^*F_v$ force measuring formula (14) measures the *instantaneous density-force displacements* (of increase/ of decrease): for a limited non-inertial accelerating displaced mass $+\underline{m}$. And which this same mass $+\underline{m}$ is being "instantaneously impacted", or "continuously impacting" by the dynamic actions of a (classical) limited force: $+F_v$. And these "dynamic actions" are happening, instantaneously, within a time interval $\underline{t}_b$, from (0 to 1) of one second. Or are happening continuously, within a larger time interval.

And although, we can always discuss and consider the "force", of a limited non-inertial accelerating displaced mass $+\underline{m}$, in free vacuum—by using the limited (classical) Newtonian formula: $+\underline{m}\dot{v}$, for force.

But according to my new mechanical theory, we must make use of a limiting non-inertial factor: $(+1-(\underline{u}/\underline{c})^2)^{+\frac{1}{2}}$, which is divided into the limited (classical) Newtonian measuring formula: $+\underline{m}\dot{v}$; ie., we have: $(+\underline{m}\dot{v})/(+1-(\underline{\dot{u}}/\underline{c})^2)^{+\frac{1}{2}}$. See formula (11). And then by our use of this limiting non-inertial factor, the new empirical conception of: instantaneous measured *acceleration-force displacements* (of decrease/of increase)—for a limited non-inertial accelerating displaced mass $+\underline{m}$, comes into play, as the new dynamic force measuring formula (11) indicates.

And we must see that this new empirical conception of: "instantaneous measured acceleration-force displacement—(of increase/of decrease)" is empirically explained as follows: if $(+1-(\underline{\dot{u}}/\underline{c})^2)^{+\frac{1}{2}}$ is less than a term: $+\underline{k}$, in the inequality (9), (see introduction). And then, according to the formula (11), ie., $(+\underline{m}\dot{v})/(+1-(\underline{\dot{u}}/\underline{c})^2)^{+\frac{1}{2}} = +q$, and thus: $+\underline{m}\dot{v} < +q$. And which this inequality (9) signifies, empirically, that formula (11), measures the increase displacement, in regards to its: *instantaneous acceleration-force displacements*. As is experienced by a non-inertial accelerating displaced mass $+\underline{m}$. As is also caused by the dynamic actions of a limited (classical) force. And so on.

We also must note that instead of the *instantaneous acceleration-force displacement*, we could use the conception of: *continuous acceleration-force displacement*, within a larger time interval.

And also, according to formula (11), if the limiting non-inertial factor: $(+1-(\underline{\dot{u}}/\underline{c})^2)^{+\frac{1}{2}}$, is greater than the term: $+\underline{k}$, in the inequality (9). And then, according to the formula (11), $(+\underline{m\dot{v}})/(+1-(\underline{\dot{u}}/\underline{c})^2)^{+\frac{1}{2}} = +q$, and thus: $+\underline{m\dot{v}} > +q$. And which this inequality (9), signifies, empirically, that formula (11) measures the decrease displacements, in regards to its: *instantaneous acceleration-force displacements*. As is experienced by a non-inertial accelerating displaced mass $+\underline{m}$. As is also caused by the dynamic actions of a limited (classical) force. And so on.

And also, according to formula (11), if the limiting non-inertial factor: $(+1-(\underline{\dot{u}}/\underline{c})^2)^{+\frac{1}{2}}$ is greater than a term: $+\underline{k}$, in the inequality (9), (see introduction). And then according to formula (12), ie., $(+\underline{m\dot{v}})(+1-(\underline{\dot{u}}/\underline{c})^2)^{+\frac{1}{2}} = +s$; and thus: $+\underline{m\dot{v}} < +s$. And which this inequality signifies, empirically, that formula (12) measures the increase displacements, in regards to its *instantaneous mass-force displacements*. As is experienced by a non-inertial accelerating displaced mass: $+\underline{m}$—as is caused by the dynamic actions of a limited (classical) force. And so on.

And also, according to formula (12), if the limiting inertial factor: is less than a term $+\underline{k}$, in the inequality (9), (see introduction). And then according to formula (12), ie., $(+\underline{m\dot{v}})(+1-(\underline{\dot{u}}/\underline{c})^2)^{+\frac{1}{2}} = +T$; and thus: $+\underline{m\dot{v}} > +T$. And which this inequality signifies, empirically, that formula (12) measures the decrease displacements, in regards to its *instantaneous mass-force displacements*. As is experienced by a non-inertial accelerating displaced mass $+\underline{m}$, as caused by the dynamic actions of a limited force. And so on.

I have used the (classical) Newtonian force measuring formula: $(+\underline{m\dot{v}})$, in the two limited force measuring formulas (11), and (12). And according to classical physics, the empirical conception of "force" is given by this (classical) Newtonian force measuring formula: $(+\underline{m}\dot{v})$. And perhaps, we can call this original (classical) Newtonian force measuring formula: $(+\underline{m}\dot{v})$, as "mass-force".

And it is in this sense, we can use the (classical) Newtonian force measuring formula: $(+\underline{m}\dot{v})$, and derive directly from it the two other new (classical) measuring formulas: $(\underline{V}\dot{v})$, $(+\underline{d}\dot{v})$. And where the term: $+\underline{V}$ is scalar volume, and where the term: $+\underline{d}$, is scalar density—with respect to the a scalar mass term: $+\underline{m}$. And where I have used these two other new (classical) force measuring formulas: $(+\underline{v}\dot{v})$, and $(+\underline{d}\dot{v})$, in the two other new limited force measuring formulas (13) and (14).

And if the original (classical) Newtonian force measuring formula: $(+\underline{m}\dot{v})$, can be called the: "mass-force" formula, and then, the other new (classical) force measuring formula: $(+\underline{V}\dot{v})$, can be called: "volume-force". And while the one other new (classical) force measuring formula: $(+\underline{d}\dot{v})$, can be called: "density-force".

These two new (classical) force measuring formulas: $(+\underline{V}\dot{v})$, and $(+\underline{d}\dot{v})$, can take their own natural place as being newly derived (classical) force formulas—with respect to the original (classical) Newtonian force formula: $(+\underline{m}\dot{v})$, which they are derived from.

And now, as based upon this preceding physical analysis, we can now always discuss and consider the "volume-force"—with respect to the volume $+\underline{V}$ aspect of a limited non-inertial accelerating displaced mass $+\underline{m}$, in free vacuum. That is, by using the new other derived limited (classical) force measuring formula: $(+\underline{V}\dot{v})$, as "volume-force".

And also, as based upon this same preceding physical analysis, we can now discuss and consider the "density-force", with respect to the density $+\underline{d}$ aspect of a limited non-inertial accelerating displaced mass $+\underline{m}$, in free vacuum. That is, by using the new other derived limited (classical) force measuring formula: $(+\underline{d}\dot{v})$, as "density-force".

But according to my new mechanical theory, we must make use of a limiting non-inertial factor: $(+1-(\underline{\dot{u}}/\underline{c})^2)^{+\frac{1}{2}}$ is divided into the new derived limited (classical) volume-force measuring formula: $(+\underline{V}\dot{v})$, ie., $(+\underline{V}\dot{v})/(+1-(\underline{\dot{u}}/\underline{c})^2)^{+\frac{1}{2}}$. (See formula (13)). And then by our use of this limiting non-inertial factor, then the empirical conception of: "instantaneous measured volume-force displacements (of decrease/of increase)—for a limited non-inertial accelerating displaced mass $+\underline{m}$ comes into play. As the new dynamic force measuring formula: (13) indicates.

And we must see that this new empirical conception of: "instantaneous measured volume-force displacements (of decrease/of increase" is empirically explained as follows: if the limiting non-inertial factor: $(+1-(\underline{\dot{u}}/\underline{c})^2)^{+\frac{1}{2}}$ is greater than a term: $+\underline{k}$, in the inequality (9), (see introduction). And thus, according to formula: (13), ie., $(+\underline{V}\dot{v})/(+1-(\underline{\dot{u}}/\underline{c})^2)^{+\frac{1}{2}} = +U$; and thus: $+\underline{V}\dot{v} > +U$. And this inequality signifies, empirically, that formula (13) measures the decrease displacements, in regards to its *instantaneous volume-force displacements*—as it is experienced by a non-inertial accelerating displaced mass $+\underline{m}$, as caused by the dynamic actions of a limited force. And so on.

And also, according to formula (13), if the limiting non-inertial factor: $(+1-(\underline{\dot{u}}/\underline{c})^2)^{+\frac{1}{2}}$ is less that a term: $+\underline{k}$, in the inequality (9), (see introduction). And thus, according to formula (13), ie., $(+\underline{V}\dot{v})/(+1-(\underline{\dot{u}}/\underline{c})^2)^{+\frac{1}{2}} = +W$; and where: $(+\underline{V}\dot{v} < +W)$. And which this inequality signifies, empirically, that formula (13) measures the increase displacements—in regards to its *instantaneous volume-force displacements*. And as is experienced by a non-inertial accelerating displaced mass $+\underline{m}$—as caused by the dynamic actions of a limited force. And so on.

And now, according to my new mechanical theory, we must make use of a limiting non-inertial factor: $(+1-(\underline{\dot{u}}/\underline{c})^2)^{+\frac{1}{2}}$, which is multiplied into the new derived limited (classical) density-force formula: $+\underline{d}\dot{v}$, ie., $(+\underline{d}\dot{v})(+1-(\underline{\dot{u}}/\underline{c})^2)^{+\frac{1}{2}}$. (See introduction). And then by our use of this limiting non-inertial factor, the new empirical conception of: "instantaneous measured density-force displacements (of increase/ of decrease)"—for a limited non-inertial accelerating displaced mass $+\underline{m}$, comes into play. As the new dynamic force measuring formula (14) indicates.

And we must also see that this new empirical conception of: "instantaneous measured density-force displacements (of increase/ of decrease)", is empirically explained as follows: if the limiting non-inertial factor: $(+1-(\underline{\dot{u}}/\underline{c})^2)^{+\frac{1}{2}}$ is greater than a term: $+\underline{k}$, in the inequality (9), (see introduction). And then according to formula (14): $(+\underline{\dot{v}})(+1-(\underline{\dot{u}}/\underline{c})^2)^{+\frac{1}{2}} = +X$; and thus: $+\underline{\dot{v}} < +X$. And which this inequality signifies, empirically, that formula (14) measures the increase displacement—in regards to its *instantaneous density-force displacements*. As is experienced by a non-inertial accelerating displaced mass $+\underline{m}$; as caused by the dynamic actions of a limited force. And so on.

And also, according to formula: (14), if the limiting non-inertial factor: $(+1-(\underline{\dot{u}}/\underline{c})^2)^{+\frac{1}{2}}$ is less than a term: $+\underline{k}$, in the inequality (9), (see introduction). And then according to the formula (14), ie., $(+\underline{\dot{v}})(+1-(\underline{\dot{u}}/\underline{c})^2)^{+\frac{1}{2}} = +y$, and thus: $+\underline{\dot{v}} > +y$. And which this inequality signifies, empirically, that formula (14) measures the decrease displacements—in regards to its *instantaneous* density-force displacements—as is experienced by a non-inertial accelerating displaced mass $+\underline{m}$; as caused by the dynamic actions of a limited force. And so on.

And these two new dynamic force measuring formulas (11) and (12) goes hand in hand with each other. Since, there is *no instantaneous acceleration-force displacements* (*of increase/of decrease*): for a limited non-inertial accelerating displaced mass $+\underline{m}$, without there being an *instantaneous mass-force displacements* (*of increase/of decrease*), for this same limited non-inertial accelerating displaced mass $+\underline{m}$.

And these other two new dynamic force measuring formulas: (13) and (14), also goes hand in hand with each other. Since, there is no "volume-force" displacements (of decrease/of increase), without there being a "density-force" displacements (of increase/of decrease)—for this same limited non-inertial accelerating displaced mass $+\underline{m}$.

Since, if there is a "acceleration-force" displacements "of increase", for a limited non-inertial acceleration displaced mass $+\underline{m}$, this implies a "mass-force" displacements (of increase), for this same limited non-inertial accelerating displaced mass +m.

And since, if there is a "volume-force" displacements (of decrease), for a limited non-inertial accelerating displaced mass +m, this implies a "density-force" displacement (of increase), for this same limited non-inertial accelerating displaced mass +m.

And we can now show the dynamic force relations between the four dynamic force formulas: (11), (12), (13), and (14), as follows:

And which we can put: is that a limited non-inertial accelerating displaced mass +m is experiencing an instantaneous increase in its *acceleration-force displacements*. As a result of the same limited non-inertial accelerating displaced mass +m of an instantaneous decrease in its non-inertial *volume-force displacements*. And then, this same limited non-inertial accelerating displaced mass +m, is now experiencing and instantaneous increase in its *mass-force displacements*. As a result of the same limited non-inertial accelerating displaced mass +m, is experiencing an instantaneous increase in its non-inertial *density-force displacements*.

And this same limited non-inertial accelerating displaced mass +m is not acquiring extra mass from some kind of external sources. But only that its non-inertial *mass-force displacements* is being instantaneously increased. As a result of the instantaneous increase in its non-inertial *density-force displacements*. And also as a result of the instantaneous decrease in its non-inertial *volume-force displacements*.

Whereby, when a limited non-inertial accelerating displaced mass +m, is experiencing an instantaneous increase in its non-inertial *acceleration-force displacements*. As a result of the same limited non-inertial accelerating displaced mass +m is experiencing an instantaneous decrease in its non-inertial *volume-force displacements*.

As both of these two above physical events are caused by the beginning of an "instantaneous implosions" of this same real limited non-inertial accelerating displaced *mass +m into itself.* Instantaneously, within a time interval t_b, from (0 to 1), of one second; or continuously, within, or at the end of a larger time interval.

And which the beginning of these "instantaneous implosions" are caused by the intense dynamic actions of a (classical) limited force: $+F_v$—"instantaneously impacting" upon this same real and limited non-inertial accelerating displaced mass $+\underline{m}$: instantaneously, within a time interval $\underline{t}_b$, from (0 to $\underline{1}$), of one second; or continuously, within, or at the end of a larger time interval.

And whereby, when a limited non-inertial accelerating displaced mass $+\underline{m}$ is experiencing an instantaneous increase in its non-inertial *mass-force displacements*. As a result of the same limited non-inertial accelerating displaced mass $+\underline{m}$ is experiencing an instantaneous increase in its non-inertial *density-force displacements*.

As both of these two above physical results are caused by the beginning of an "instantaneous implosions" of this same real limited non-inertial accelerating displaced *mass $+\underline{m}$ into itself*. Instantaneously, within a time interval $\underline{t}_b$, from (0 to $\underline{1}$), of one second; or continuously, within a larger time interval.

And again, the beginning of these "instantaneous implosions" are caused by the intense dynamic actions of a (classical) limited force: $+F_v$—"instantaneously impacting" upon this same real limited non-inertial accelerating displaced mass $+\underline{m}$. Instantaneously, within a time interval $\underline{t}_b$, from (0 to $\underline{1}$), of one second; or continuously, within a larger time interval.

And since, if there is a "acceleration-force" displacement (of increase), for a limited non-inertial accelerating displaced mass $+\underline{m}$, this implies a "mass-force" displacement (of increase), for this same real limited non-inertial accelerating displaced mass $+\underline{m}$.

And also, if there is a "volume-force" displacements (of increase), for a limited non-inertial accelerating displaced mass $+\underline{m}$—this implies a "density-force" displacements (of decrease), for this same mass $+\underline{m}$.

And we can show the other dynamic force relations between the four dynamic force measuring formulas: (11), (12), (13), and (14), as follows:

And which we can put: is that a limited non-inertial acceleration displaced mass $+\underline{m}$ is experiencing an instantaneous "decrease" in its *acceleration-force displacements*. And as a result of the same mass $+\underline{m}$ is experiencing an instantaneous "increase" in its non-inertial *volume-force displacements.* And then, this same mass +m̲ is also now experiencing an instantaneous "decrease" in its non-inertial *mass-force displacements.* And as a result, of the same mass +m̲ is now experiencing an instantaneous "decrease" in its non-inertial *density-force displacements.*

Whereby, when a limited non-inertial accelerating displaced mass +m̲ is experiencing an instantaneous increase in its non-inertial *acceleration-force displacements*—as a result of the same limited non-inertial accelerating displaced mass +m̲ is experiencing an instantaneous increase in its non-inertial *volume-force displacements.*

As both of these two physical events are caused by the beginning of an "instantaneous explosions" of this same mass +m̲ *outwards from itself*—instantaneously, within a time interval t_b, from (0 to 1), of one second; or continuously, within a larger time interval.

And which the beginning of these "instantaneous explosions" are caused by the intense dynamic actions of a (classical) limited force: +F_v—"instantaneously impacting" upon this same real limited mass +m̲. Instantaneously, within a time interval t_b, from (0 to 1), of one second. Or for this same real limited force: F_v is "continuously impacting" upon this same real limited mass +m̲, continuously, within a larger time interval.

And whereby, when a limited non-inertial accelerating displaced mass +m̲ is experiencing an instantaneous decrease in its non-inertial *mass-force displacements.* As a result of the same real limited mass +m̲ is experiencing an instantaneous decrease in its non-inertial *density-force displacements.*

As both of these two physical events are caused by the beginning of an "instantaneous explosions" of this same real limited non-inertial accelerating displaced mass +m̲ *outwards from itself.* Instantaneously, within a time interval t_b, from (0 to 1), of one second; or continuously, within, or at the end of a larger time interval.

And which the beginning of these "instantaneous explosions" are caused by the intense dynamic actions of a (classical) limited force: +F_v—"instantaneously impacting" upon this same real limited mass +m̲. Instantaneously, within a time interval t_b, from (0 to 1), of one second. Or these "continuous explosions" for a mass +m̲ are caused by the dynamic actions of a (classical) limited force: F_v—"continuously impacting" upon this same real limited mass $+\underline{m}$. Continuously, within, or at the end of a larger time interval.

And these two new empirical conditions of: "instantaneous implosions/ instantaneous explosions" are caused by the intense dynamic actions of a (classical) limited force: F_v. Which is "instantaneously impacting" upon this same real limited mass $+\underline{m}$. And these "intense dynamic actions" are happening, instantaneously, within a time interval $\underline{t}_b$, from (0 to 1) of one second; or a larger time interval.

And also the two new empirical conditions of: "continuous implosions/ continuous explosions" are caused by the intense dynamic actions of a (classical) limited force: $+F_v$. Which are "continuously impacting" upon this same limited mass $+\underline{m}$. And these "intense dynamic actions" are happening, continuously, within, or at the end of a larger time interval.

I will now offer a new limited dynamic acceleration measuring formula which will measure separately the "acceleration" aspect of the "acceleration-force" displacements of formula (11). And I will also offer a new dynamic "mass" measuring formula which measures separately the "mass" aspect of the "mass-force" displacement of formula (12).

And I will also offer a new limited dynamic volume measuring formula which will measure separately the "volume" aspect of the "volume-force" displacement of formula (13). And I will also offer a new limited dynamic density formula which will measure separately the "density" aspect of the "density-force" displacement of formula (14).

The instantaneous, or continuous, limited non-inertial "acceleration $+\dot{v}$ displacement (of increase/of decrease), for a limited non-inertial accelerating displaced mass $+\underline{m}$, is measured and given by the new measuring formula: $(+\dot{v})/(+1-(\underline{\dot{u}}/\underline{c})^2)^{+\frac{1}{2}}$. And where the mass term: $+\underline{m}$ is cancelled out of formula (11).

And the instantaneous limited non-inertial accelerating displaced mass $+m$ displacement (of increase/of decrease), for a limited displaced mass $+\underline{m}$ is measured and given by the new measuring formula:

$(+\underline{m})(+1-(\underline{\dot{u}}/\underline{c})^2)^{+\frac{1}{2}}$. And where the uniform limited non-inertial accelerating term: $+\dot{v}$ is cancelled out of formula (12).

And the instantaneous limited non-inertial accelerating "volume $\underline{V}$" displacements (of decrease/of increase), for a limited non-inertial accelerating displaced mass $+\underline{m}$, is measured and given by the new measuring formula: $(+\underline{V})/(+1-(\underline{\dot{u}}/\underline{c})^2)^{+\frac{1}{2}}$. And where the uniform limited non-inertial acceleration term: $+\dot{v}$, and the uniform density term: $+\underline{d}$, as both of these two terms are cancelled out of formula (11).

And the instantaneous limited non-inertial accelerating "density $\underline{d}$" displacements (of increase/of decrease), for a limited non-inertial accelerating displaced mass $+\underline{m}$, is measured and given by the new measuring formula: $(+\underline{d})(+1-(\underline{\dot{u}}/\underline{c})^2)^{+\frac{1}{2}}$. And where the uniform limited non-inertial accelerating term: $+\dot{v}$, and the uniform scalar volume term: $+\underline{V}$, as both of these terms are cancelled out of formula (12).

And I also want to note, that the same kind of analysis, and physical interpretations, I had previously given to the four new limited dynamic force measuring formulas: (11), (12), and (13), and (14)—also similarity applies to these above four new limited dynamic measuring formulas.

And now, according to the four new limited dynamic: $+F_v$ force measuring formulas: (11), (12), (13), and (14), which they measure the extreme "instantaneous implosion effects", of a real and limited non-inertial accelerating displaced *mass* $+\underline{m}$ *into itself*, which it will experience. And which these "effects" are caused by the beginning of the intense dynamic actions of a limited (classical) force: $+F_v$. Which instantaneously impacting upon this real and limited non-inertial accelerating displaced mass $+\underline{m}$, instantaneously, within a time interval $\underline{t}_b$, from (0 to 1) of one second, or continuously, within, or at the end of a larger time interval.

And also, according to these same four limited dynamic $+F_v$ force measuring formulas: (11), (12), (13), and (14), which they also measure the extreme "instantaneous explosion effects", of a real and limited non-inertial accelerating *displaced mass* $+\underline{m}$ *outwards from itself*, which it will experience. And which these new "effects" are caused by the beginning of the intense dynamic actions of a (classical) limited force: F_v, "instantaneously impacting" upon this same real and limited non-inertial accelerating displaced mass $+\underline{m}$—instantaneously, within a time interval $\underline{t}_b$, from (0 to 1), of one second.

And also according to these same four new limited dynamic $+F_v$, force measuring formulas (11), (12), (13), and (14), which they also measure the extreme "continuous explosion effects" of a real and limited non-inertial accelerating *displaced mass* $+\underline{m}$, *outwards from itself*, which it will experience. And which they "effects" are caused by the beginning of the intense dynamic actions of a (classical) limited force: $+F_v$, "continuously impacting" upon this same real and limited non-inertial accelerating displaced mass $+\underline{m}$; continuously, within a larger time interval.

And the problem of which one of these four new empirical conditions of: "instantaneous implosions/instantaneous explosion", (effects), and "continuous implosions/ continuous explosions, (effects), that a real and limited non-inertial accelerating displaced mass $+\underline{m}$ will experience, depends on how a force: $+F_v$ is used. And perhaps upon the kind of mass $+\underline{m}$ that is used.

And it may be the case that these four new empirical conditions of: "instantaneous implosion effects/instantaneous explosion effects", and: "continuous implosion effects/continuous explosion effects" will occur simultaneously, for one or the other of these two "effects". With respect to a real and limited non-inertial displaced mass $+\underline{m}$. And again this scenario depends upon the force-magnitudes of this same (classical) limited force: $+F_v$. And also upon the kind of mass $+\underline{m}$ used.

And we must note, that the beginning of these unique "dynamic actions" that this real and limited non-inertial accelerating displaced mass $+\underline{m}$, are happening, instantaneously, within a time interval $\underline{t}_b$, for (0 to 1), of one second. Or are happening, "continuously, within, or at the end of a larger time interval".

But we must also note, that these four new empirical conditions of: "instantaneous implosion effects/instantaneous explosion effects", and of: "continuous implosion effects/continuous explosion effects", for all these four conditions are caused by a limited force: $+F_v$. And all four of these conditions must result in extreme "thermo dynamic effects", that this same real and limited non-inertial accelerating displaced mass $+\underline{m}$ will experience.

And some of these extreme: "thermo dynamic effects" are "intense and incredible electro-magnetic radiations effects". And: "incredible and extremely hot plasmatic effects" And again, of "intense implosions/ intense explosions" of this same real and limited non-inertial accelerating displaced mass $+\underline{m}$ will experience. An "intense implosion/intense explosion" that a force: $+F_v$ will cause by its "impacting" upon this same mass $+\underline{m}$.

And in the introduction of this book, I have set: $(+\underline{m}c = 0)$, and even: $(+\underline{m}c = 0)$. And now, it is easily seen that this real mass $+\underline{m}$, as indicated in these two new (classical) limit momentum measuring formulas, would have been totally converted into "intense dynamic effects", and into "intense kinematic effects". And as well as this same real and limited mass $+\underline{m}$, as also indicated in these two new (classical) limit momentum measuring formulas, would have been totally converted into "intense thermo dynamic and kinematic effects".

And which the beginning of these "intense effects" are happening to this same real and limited mass $+\underline{m}$ are caused by the former dynamic actions of a (classical) limited force: $+F_v$. Instantaneously, within a time interval $\underline{t}_b$, from (0 to 1), of one second. Or these "intense effects" are happening to this real and limited mass $+\underline{m}$, by the dynamic actions, of a (classical) limited force: $+F_v$—continuously, within, or a the end of a larger time interval.

And which the end of these "intense effects" are happening to this same real and limited mass $+\underline{m}$ (in the two above scenarios) are due, or are caused by the end of the instantaneous discontinuance of this same (classical) limited force: $+F_v$. Instantaneously, when a larger time interval equals one second, or greater than one second.

And this empirical scenario of where $(+\underline{m}c = 0)$, and $(+\underline{m}c = 0)$, is very different than Albert Einstein, and the many proponents of his special relativity theory, when they put: $(+mc > 0)$, and $(+\underline{m}c > 0)$. And which these kinds of conclusions on their part are based directly upon their total acceptance of the original Newtonian first law of motion. And where they consider all kinds of inertial velocity displacement states, or inertial speed displacement states as being freely given by nature.

And which they then place the limited inertial velocity displacement terms: $+\underline{u}$, $+v$, $+\underline{c}$, and $+c$, onto real and limited mass term: $+\underline{m}$. And it is in this case they then do not take into account of how any real and limited mass $+\underline{m}$ has achieved its limited inertial scalar speed displacements: $+\underline{u}$, and $+\underline{c}$—or has achieved its limited inertial vector velocity displacements: $+v$, and $+c$.

And it is because of the new revised Newtonian first law of motion (4), and also by our use of the two new empirical conceptions of "applied limited forces", and of the "instantaneous discontinuance" of these same limited forces. We then must now set: $(+\underline{mc} = 0)$, and: $(+\underline{mc} = 0)$. Where the term: $+c$ is the derived constant inertial vector velocity limit term. And where the term: $+\underline{c}$, is the measured constant and inertial speed limit of light, in free vacuum. And where both of these two terms are to be measured with respect to a basic time interval of one second.

Albert Einstein's special relativistic mass increase measuring formula: $(+\underline{m}_o / (+1-(\underline{u}/\underline{c})^2)^{+\frac{1}{2}} = +\underline{m}$, purports to measure the infinite mass increase, ie., $(+\underline{m} \to +\infty$. As will be experienced by a limited and real inertial velocity displaced mass $+\underline{m}$, when the limiting factor: $(+1-(\underline{u}/\underline{c})^2)^{+\frac{1}{2}}$ tends towards a zero value (0).

And Albert Einstein's special relativistic Newtonian force measuring formula: $(+\underline{m}_o\dot{v}) / (+1-(\underline{u}/\underline{c})^2)^{+\frac{1}{2}}$. and which also purports to measure the infinite Newtonian force increase, ie., $(+\underline{m}\dot{v} \to +\infty$. As will be experienced by a limited and real non-inertial accelerating displaced mass $+\underline{m}$ when the limiting inertial factor: $(+1-(\underline{u}/\underline{c})^2)^{+\frac{1}{2}}$ tends towards a zero value (0).

And these two conclusions are based directly upon my reinterpretations of Albert Einstein's special relativistic mass increasing measuring formulas, as we have seen.

But I have rejected Albert Einstein's original special relativistic mass increase measuring formula, (see introduction of this book about this matter). And then, I have completely rejected Einstein's special relativistic Newtonian force measuring formula.[8] And as both of his two measuring formulas are false measuring formulas, according to my new mechanical theory.

And nor I have used an inertial rest mass term: $+\underline{m}_o$, (in direct contrast to Albert Einstein's special relativity theory). This is the case in regards to all of the four new limited dynamic force: $+F_v$ force measuring formulas (11), (12), (13), and (14), and so on. And nor in any of the four new limit dynamic: $+F_c$ lower-force measuring formulas (15), (16), (17), and (18). (see the end of this section about this point). And the empirical reason why I have not used an inertial rest mass term: $+m$, in any of my dynamic force measuring formulas in this section, or even in this book—is based directly upon my use and acceptance of a diagram (8), (please see introduction of this book about this matter).

In fact, as we have seen, Albert Einstein's special relativistic force measuring formula: $(+\underline{m}_o \dot{v}) / (+1 - (\underline{u}/\underline{c})^2)^{+\frac{1}{2}} = +\underline{m}\dot{v}$, (with respect to the limiting inertial factor: $(+1 - (\underline{u}/\underline{c})^2)^{+\frac{1}{2}}$ is a false measuring formula).

Since, when this limiting inertial factor is numerically analyzed, and physically interpreted; and then Albert Einstein's special relativistic Newtonian force measuring formula will give a force "decrease displacement". That is, when this same limiting inertial factor is greater than a term: $+\underline{k}$. And which it tends towards a value $+\underline{3}$. (Having the unit of measurement of 10^4 kilometers per one second, see inequality (9), in the introduction).

And thus, Albert Einstein's special relativistic Newtonian force measuring formula—will give a force "increase displacements" which tends towards infinite force increase displacement. That is, when this same limiting inertial factor is less than a term; $+\underline{k}$, and which the same factor tends towards a zero value: (0). And which this factor has the unit and value of: $(+1 - (\underline{u}/\underline{c})^2)^{+\frac{1}{2}}$.

And it is because of these facts, that there are no real empirical interpretations I can give to Albert Einstein's special relativistic Newtonian measuring formula. This is because, empirically, it is a false measuring formula, according to my new mechanical theory.

And now, in concluding this section, as I have offered the four new limited +Fv force measuring formulas (11), (12), (13), and (14)—I will now offer the four new limit dynamic +F$_c$ lower-force measuring formulas. And we have, in regards to a mass term: +m

$$(+\underline{m}\dot{c})/(+1-(\dot{\underline{c}}/\underline{c})^2)^{+\frac{1}{2}} = (+F_c' = 0) \tag{15}$$

and:

$$(+\underline{m}\dot{c})(+1-(\dot{\underline{c}}/\underline{c})^2)^{+\frac{1}{2}} = (+'F_c = 0) \tag{16}$$

And we also now have, (in regards to a volume term: +V, and of a density term: +d, with respect to a mass term: +m; the other two new limit dynamic +F$_c$ lower-force measuring formulas:

$$(+\underline{V}\dot{c})/(+1-(\dot{\underline{c}}/\underline{c})^2)^{+\frac{1}{2}} = (+F_C^* = 0) \tag{17}$$

and:

$$(+\underline{d}\dot{c})(+1-(\dot{\underline{c}}/\underline{c})^2)^{+\frac{1}{2}} = (+^*F_C = 0) \tag{18}$$

And where $|+\dot{c}| = \dot{c} = 0$. And where the term: +V is scalar volume, and where the term: +d is scalar density. And where $(+\underline{V})(+\underline{d}) = +\underline{m}$, a scalar mass term.

And where the new limit +F$_c'$ lower-force formula (15) measures the *instantaneous limit acceleration-force displacements* (of zero decrease/of zero increase), for a mathematical non-existent displaced mass +m, of a time interval of one zero second.

And where the new limit +'F$_C$ lower-force formula (16) measures the *instantaneous limit mass-force displacements* (of zero increase/of zero decrease), for a mathematical non-existent displaced mass $+\underline{m}$, of a time interval of one zero second.

And where the new limit $+F_c^*$ lower-force formula (17) measures the *instantaneous limit volume-force displacements* (of zero decrease/of zero increase), for a mathematical non-existent displaced mass $+\underline{m}$, of a time interval of one zero second.

And where the new limit dynamic $+*F_c$ lower-force formula (18) measures the *instantaneous limit density-force displacements* (of zero increase/of zero decrease), for a mathematical non-existent displaced mass $+\underline{m}$, of a time interval of one zero second.

With respect to the mathematical first derivative of the limit vector velocity limit term: $+c$, is: $|+\dot{c}| = (+\dot{c} = 0)$. And which must then result, mathematically, in a zero value (0), for this real and limited mass term: $+\underline{m}$. And for this same real and limited volume term: $+\underline{V}$; and for this same real and limited density term: $+\underline{d}$. In regards to these four new limit dynamic $+F_c$ lower-force measuring formulas: (15), (16), (17), and (18).

And this is the reason, mathematically, that a real and limited mass term: $+\underline{m}$, and a real and limited density term: $+\underline{d}$, and a real and limited volume term: $+\underline{V}$—are all considered to be non-existent in this mathematical sense.

That is, in this mathematical sense of taking the first derivative of the uniform vector velocity limit term: $+c$. And which is: $(|\dot{c}| = 0)$—in regards to these four new limit dynamic $+F_c$ lower-force measuring formulas: (15), (16), (17), and (18).

And these above mathematical scenarios are very much different that the real empirical scenarios of where I have set: $(+\underline{mc} = 0)$, and even: $(+\underline{mc} = 0)$. As this is based upon the intense dynamic and kinematic effects. And also base upon the intense thermo dynamic and kinematic effects that this same mass $+\underline{m}$ will experience.

I cannot offer a real and limited empirical interpretations of these four new limit $+F_c$ lower-force measuring formulas: (15), (16), (17), and (18). And the only empirical interpretations I can give to these same four new limit dynamic lower-force measuring formulas: are only of a mathematical interpretations, and not of a real empirical interpretations.

And only to preserve the form of this book, is the only reason why I have included these four new limit dynamic $+F_c$ lower-force measuring formulas in the text of this book.

But I should note, in hindsight, that these four new limit dynamic $+F_c$ lower-force measuring formulas: (15), (16), (17), and (18), can serve as a natural lower-force limits for all limited and real force measuring formulas.

THE NEW KINEMATIC MOMENTUM MEASURING FORMULAS

In this section, I will formulate the four new limited kinematic $+P_v$ momentum measuring formulas—by using the limited (classical) momentum measuring formula: $+\underline{m}v$. And by using also the formula: $+\underline{m} = (+\underline{V})(+\underline{d})$. Where the term: $+\underline{V}$ is scalar volume, and where the term: $+\underline{d}$ is scalar density.[8] And I also will be using the limiting inertial factor: $(+1-(\underline{u}/\underline{c})^2)^{+\frac{1}{2}}$. And which I have not attempted to derive this limiting inertial factor from the two postulates (5) and (6). And in this book, it must be accepted as it is.

And I will also formulate, in this section, the four new limit kinematic $+P_c$ lower-momentum measuring formula, by using a new (classical) lower-momentum measuring formula: $+\underline{m}c = 0$). Where the term: $+c$ is the constant derived vector velocity limit term. And even of our use of: $(+\underline{m}c = 0)$, where the term: $+\underline{c}$ is the constant measured inertial speed limit $\underline{c}$ of light, in free vacuum. And also[9] by our use of the formula: $(+\underline{m}) = (+\underline{V})$ $(+\underline{d})$. And also by our use of a new limiting factor: $(+1-(\underline{u}/\underline{c})^2)^{+\frac{1}{2}}$, as it is derived from:

$$\operatorname*{Lim}_{\underline{v}\to\underline{c}}(+1-(\underline{u}/\underline{c})^2)^{+\frac{1}{2}} = ((+1-(\underline{c}/\underline{c})^2)^{+\frac{1}{2}} = 0) \qquad (19)$$

We now have, (in regards to a mass term: $+\underline{m}$), the two new limited momentum measuring formulas:

$$(+\underline{m}v)/(+1-(\underline{u}/\underline{c})^2)^{+\frac{1}{2}} = +P'_v \qquad (20)$$

and:

$$(+\underline{mv})(+1-(\underline{u}/\underline{c})^2)^{+\frac{1}{2}} = +'P_v \tag{21}$$

And where $|+v| = +\underline{u}$. And where the new limited kinematic $+P'_v$ momentum measuring formula (20) measures the *instantaneous velocity-momentum displacement* (of decrease/of increase), for a limited inertial velocity displaced mass $+\underline{m}$. Instantaneously, at the end of the intense dynamic actions of a (classical) limited force: $+F_v$. That is, at the end of the instantaneous discontinuance of this same (classical) limited force: $+F_v$—instantaneously, at the end of a time interval $\underline{t}_b$, when $\underline{t}_b = 1$ second; or continuously, at the end of a larger time interval.

And where the new limited kinematic $+P_v$ momentum measuring formula (21) measures the instantaneous *mass-momentum displacement* (of increase/of decrease), for a limited inertial velocity displaced mass $+\underline{m}$. Instantaneously, at the end of the intense dynamic actions of a (classical) limited force: $+F_v$. That is, at the end of the instantaneous discontinuance of this same (classical) limited force: $+F_v$. Again, instantaneously, at the end of a time interval $\underline{t}_b$, when $\underline{t}_b = 1$ second: Or continuously, at the end of a larger time interval.

We now have: a volume term: $+\underline{V}$, and a density term: $+\underline{d}$, with respect to a mass term: $+\underline{m}$. And thus, the other new two limited kinematic momentum measuring formula:

$$(+\underline{V}v)/(+1-(\underline{u}/\underline{c})^2)^{+\frac{1}{2}} = +P_v^* \tag{22}$$

And:

$$(+\underline{d}v)(+1-(\underline{u}/\underline{c})^2)^{+\frac{1}{2}} = +{}^*P_v \tag{23}$$

And where $|+v| = +\underline{u}$. And where the new limited kinematic $+P_v^*$ measuring momentum formula (22) measures the *instantaneous volume-momentum displacement* (of decrease/of increase), for a limited inertial velocity displaced mass $+\underline{m}$. And again, instantaneously, at the end of the intense dynamic actions of a (classical) limited force: $+F_v$. And at the end of the instantaneous discontinuance of this same (classical) limited force: $+F_v$, again, instantaneously, at the end of a time interval $\underline{t}_b$, when $\underline{t}_b = 1$ second: or continuously, within, or at the end of a larger time interval.

And where the new limited kinematic $+^*P_v$ momentum measuring formula (23) measures the *instantaneous density-momentum displacement* (of increase/of decrease), for a limited inertial velocity displaced mass $+\underline{m}$, again, instantaneously, at the end of the intense dynamic actions of a (classical) limited force: $+F_v$. That is, at the end of the instantaneous discontinuance of this same (classical) limited force: $+F_v$; and again, instantaneously, at the end of a time interval $\underline{t}_b$, when $\underline{t}_b = 1$ second. Or continuously, at the end of a larger time interval.

And although, we can always discuss and consider the "momentum" of a limited inertial velocity displaced mass $+\underline{m}$, in free vacuum space and time. By using the limited (classical) formula: $+\underline{m}v$, for momentum.

But according to my new mechanical theory, we must make use of a limiting inertial factor: $(+1-(\underline{u}/\underline{c})^2)^{+\frac{1}{2}}$ which is divided into the limited (classical) momentum formula: $+\underline{m}v$, ie., $(+\underline{m}v)/(+1-(\underline{u}/\underline{c})^2)^{+\frac{1}{2}}$. (See formula (20)). And then, by our use of this limiting inertial factor, the new empirical conception of: "instantaneous measured velocity momentum" displacement (of decrease/of increase), for a limited inertial velocity displaced mass $+\underline{m}$, comes into play. As the new kinematic momentum measuring formula (20) indicates.

And we must see that the new empirical conception of: "instantaneous measured velocity-momentum displacement (of decrease/of increase)" is empirically explained as follows: if a limiting inertial factor: $(+1-(\underline{u}/\underline{c})^2)^{+\frac{1}{2}}$ is greater than a term: $+\underline{k}$, in the inequality (9), (see introduction). And then according to formula (20), ie., $(+\underline{m}v)/(+1-(\underline{u}/\underline{c})^2)^{+\frac{1}{2}} = +A$; and thus: $+\underline{m}v > +A$. And which this inequality (9) signifies, empirically, that formula (20) measures the decrease displacement, in regards to its *instantaneous velocity-momentum displacement*. As it is experienced by an inertial velocity displaced mass $+\underline{m}$. As were caused by the instantaneous discontinuance of the dynamic actions of a limited force. And so on.

And also, according to formula (20), if the limiting inertial factor: $(+1-(\underline{u}/\underline{c})^2)^{+\frac{1}{2}}$ is less than a term: $+\underline{k}$, in the inequality (9), (see introduction). And then, according to formula (20), ie., $(+\underline{mv})/(+1-(\underline{u}/\underline{c})^2)^{+\frac{1}{2}} = +B$. And thus: $+\underline{mv} < +B$. And which this inequality (9) signifies, empirically, that formula (20) measures the increase displacement, in regards to its *instantaneous velocity-momentum displacement*, as it is experienced by an inertial velocity displaced mass $+\underline{m}$. As were caused by the instantaneous discontinuance of the dynamic actions of a limited force. And so on.

And according to my new mechanical theory, we must make use of a limited inertial factor: $(+1-(\underline{u}/\underline{c})^2)^{+\frac{1}{2}}$, which is multiplied into the limited (classical) momentum formula: $+\underline{mv}$, ie., $(+\underline{mv})(+1-(\underline{u}/\underline{c})^2)^{+\frac{1}{2}}$. (See formula (21)). And then by our use of this limiting inertial factor, the new empirical conception of: "instantaneous measured mass-momentum displacement (of increase/of decrease)"—for a limited inertial velocity displaced mass $+\underline{m}$, comes into play. As the new kinematic momentum measuring formula (21) indicates.

And also we must see that this new empirical conception of: "instantaneous measured mass-momentum displacement (of increase/ of decrease)" is empirically explained as follows: if the limiting inertial factor: $(+1-(\underline{u}/\underline{c})^2)^{+\frac{1}{2}}$ is greater than a term: $+\underline{k}$, in the inequality (9), (see introduction). And then, according to the formula (21), ie., $(+\underline{mv})(+1-(\underline{u}/\underline{c})^2)^{+\frac{1}{2}} = +C$; and thus: $+\underline{mv} < +C$. And which this inequality signifies, empirically, that formula (21) measures the increase displacement, in regards to its *instantaneous mass-momentum displacement*. As it is experienced by an inertial velocity displaced mass $+\underline{m}$. As also were caused by the instantaneous discontinuance of the dynamic actions of a limited force. And so on.

And also, according to formula (21), if the limiting inertial factor: $(+1-(\underline{u}/\underline{c})^2)^{+\frac{1}{2}}$ is less than a term: $+\underline{k}$, in the inequality (9), (see introduction).

And then, according to the formula (21), ie., $(+\underline{m}v)(+1-(\underline{u}/\underline{c})^2)^{+\frac{1}{2}} = +D$. And thus: $+\underline{m}v > +D$. And which this inequality signifies, empirically, that formula (21) measures the decrease displacement—in regards to its *instantaneous mass-momentum displacement*. As is experienced by an inertial velocity displaced mass $+\underline{m}$; as were caused by the instantaneous discontinuance of the dynamic actions of a limited force. And so on.

I have used the (classical) momentum measuring formula: $(+\underline{m}v)$, in the two new limited momentum measuring formulas: (20) and (21). And according to classical physics, the empirical conception of "momentum" is given by this (classical) momentum measuring formula: $+\underline{m}v$. And perhaps, we can call this original (classical) momentum measuring formula: $+\underline{m}v$, as "mass-momentum", and also as: "velocity-momentum". Which these names depends upon how the limiting inertial factor: $(+1-(\underline{u}/\underline{c})^2)^{+\frac{1}{2}}$ is used upon the (classical) momentum formula: $+\underline{m}v$.

And it is in this sense, we can use this (classical) momentum measuring formula: $+\underline{m}v$, and derive directly from it the two other new (classical) momentum measuring formulas: $(+\underline{V}v)$, and $(+\underline{d}v)$. And where the term: $+\underline{V}$ is scalar volume, and where the term: $+\underline{d}$ is scalar density—with respect to a scalar mass term: $+\underline{m}$. And I have used these two other new (classical) momentum measuring formulas: $(+\underline{V}v)$, and $(+\underline{d}v)$, in the two other new limited momentum measuring formulas (22) and (23).

And if the original momentum formula: $(+\underline{m}v)$ be called "mass-momentum", and "velocity-momentum",—then the one other new (classical) momentum measuring formula: $(+\underline{V}v)$ can be called "volume-momentum". And while the other new (classical) momentum measuring formula: $(+\underline{d}v)$ can be called "density-momentum".

And why I have called these two other new momentum measuring formula: $(+\underline{V}v)$, and $(+\underline{d}v)$ "classical" is because they are directly derived from the original (classical) momentum measuring formula: $(+\underline{m}v)$.

These two new (classical) momentum measuring formulas: $(+\underline{V}v)$, and $(+\underline{d}v)$, can take their own natural place as being newly derived (classical) momentum measuring formulas, with respect to the original (classical) momentum measuring formula: ($+\underline{m}v$), which they are derived from.

And now, as based upon this preceding physical analysis, we can now always discuss and consider the "volume-momentum"—with respect to the volume $+\underline{V}$ aspect of a limited inertial velocity displaced mass $+\underline{m}$ in free vacuum. And thus, we have the new formula: ($+\underline{V}v$), as "volume-momentum".

And also, as based upon this preceding physical analysis, we can now always discuss and consider the "density-momentum"—with respect to the $+\underline{d}$ aspect of a limited inertial velocity displaced mass $+\underline{m}$, in free vacuum. And thus, we have the new formula: ($+\underline{d}v$), as "density-momentum".

But according to my new mechanical theory, we must make use of a limiting inertial factor: $(+1-(\underline{u}/\underline{c})^2)^{+\frac{1}{2}}$, which is divided into the new derived limited (classical) volume-momentum measuring formula: ($+\underline{V}v$), ie., $(+\underline{V}v)/(+1-(\underline{u}/\underline{c})^2)^{+\frac{1}{2}}$. (See formula (22)). And then by our use of this limiting inertial factor, the new empirical conception of: "instantaneous measured "volume-momentum" displacements (of decrease/of increase), for a limited inertial velocity displaced mass $+\underline{m}$, comes into play, as the new kinematic momentum measuring formula (22) indicates.

And we must see that the new empirical conception of: "instantaneous measured volume-momentum displacements (of decrease/of increase)" is empirically explained as follows: If the limiting inertial factor: $(+1-(\underline{u}/\underline{c})^2)^{+\frac{1}{2}}$ is greater than a term: $+\underline{k}$, in the inequality (9), (see introduction). And then according to formula (22), ie., $(+\underline{V}v)/(+1-(\underline{u}/\underline{c})^2)^{+\frac{1}{2}} = +g$, and thus: $+\underline{V}v > +g$. And which this inequality signifies, empirically, that formula (22) measures the decrease displacement—in regards to its instantaneous *volume-momentum displacements*. As it is experienced by an inertial velocity displaced mass $+\underline{m}$. As were caused by the instantaneous discontinuance of the dynamic actions of a limited force. And so on.

And according to formula (22), if the limiting inertial factor: $(+1-(\underline{u}/\underline{c})^2)^{+\frac{1}{2}}$ is less than a term: $+\underline{k}$, in the inequality (9), (see introduction). And then, according to formula (22), ie., $(+\underline{V}v)/(+1-(\underline{u}/\underline{c})^2)^{+\frac{1}{2}} = +H$.

And thus, $+\underline{V}v < +H$. And which this inequality signifies, empirically, that formula (22) measures the increase displacements. In regards to its *instantaneous volume-momentum displacement*. As is experienced by an inertial velocity displaced mass $+\underline{m}$. As were caused by the instantaneous discontinuance of the dynamic actions of a limited force, and so on.

And now according to my new mechanical theory, we must make use of a limiting inertial factor: $(+1-(\underline{u}/\underline{c})^2)^{+\frac{1}{2}}$, which is multiplied into the new derived limited (classical) density-momentum formula: $(+\underline{d}v)$, ie., $(+\underline{d}v)(+1-(\underline{u}/\underline{c})^2)^{+\frac{1}{2}}$. (See formula (23)). And then by our use of this limiting inertial factor, the new conception of: "instantaneous measured density-momentum displacement (of increase/of decrease)". In regards to a limited inertial velocity displaced mass $+\underline{m}$, comes into play. As the new kinematic momentum measuring formula (23) indicates.

And we must also see that this new empirical conception of: "instantaneous measured density-momentum displacement (of increase/of decrease)" is empirically explained as follows: if a limiting inertial factor: $(+1-(\underline{u}/\underline{c})^2)^{+\frac{1}{2}}$ is greater than a term; $+\underline{k}$, in the inequality (9), (see introduction). And then according to formula (23) ie., $(+\underline{d}v)(+1-(\underline{u}/\underline{c})^2)^{+\frac{1}{2}} = +I$. And thus $+\underline{d}v < +I$. And which this inequality signifies, empirically, that formula (23) measures the increase displacement—in regards to its *instantaneous density-momentum displacement*. As is experienced by an inertial velocity displaced mass $+\underline{m}$. As were caused by the instantaneous discontinuance of the dynamic actions of a limited force. And so on.

And also, according to formula (23), if the limiting inertial factor: $(+1-(\underline{u}/\underline{c})^2)^{+\frac{1}{2}}$ is less than a term: $+\underline{k}$, in the inequality (9) (see introduction). And then, according to formula (23), ie., $(+\underline{d}v)(+1-(\underline{u}/\underline{c})^2)^{+\frac{1}{2}} = +J$. And thus: $+\underline{d}v > +J$. And which this inequality signifies, empirically, that formula (23) measures the decrease displacement, in regards to its *instantaneous density-momentum displacement*. As it is experienced by an inertial velocity displaced mass $+\underline{m}$—as were caused by the instantaneous discontinuance of the dynamic actions of a limited force. And so on.

And these two new kinematic momentum formulas: (20), and (21), goes hand in hand with each other. Since, there is no "velocity-momentum" displacement (of decrease/of increase)—for a limited inertial velocity displaced mass +m: without there being a "mass-momentum" displacement (of increase/ of decrease)—for this same limited inertial velocity displaced mass +m.

And these other two new kinematic momentum measuring formulas: (22) and (23), also goes hand in hand with each other. Since, there is no "volume-momentum" displacement (of decrease/of increase)—for this limited inertial velocity displaced mass +m—without there being also a "density-momentum" displacement (of increase/of decrease): for this same limited inertial velocity displaced mass +m.

Since, if there is a "velocity-momentum" displacement of "decrease, for a limited inertial velocity displaced mass +m—this implies a "density-momentum" displacement of "increase"—for this same limited inertial velocity displaced mass +m.

And we can now show the kinematic momentum relations between the four new kinematic momentum measuring formulas: (20), (21), (23), and (24), as follows:

And which we can put: is that a limited inertial velocity displaced mass +m is experiencing an instantaneous "decrease" in its inertial *velocity-momentum displacement*. As a result of the same limited inertial velocity displaced mass +m, is experiencing an instantaneous "decrease" in its inertial *volume-momentum displacement*. And then, this same limited inertial velocity displaced mass +m: is now experiencing an instantaneous "increase" in its inertial *mass-momentum displacement*. As a result of the same limited inertial velocity displaced mass +m: is experiencing an instantaneous "increase" in it *density-momentum displacements*.

And this same limited inertial velocity displaced mass +m is not acquiring extra mass from some kind of external sources. But only that its inertial *mass-momentum displacement* is being "instantaneously increased". And also as a result of the "instantaneous decrease" in its *volume-momentum displacement*.

And both of these two new above physical events are due, or are caused by the end of an "instantaneous implosions" of this same limited inertial velocity displaced *mass* +m *into itself.* Instantaneously, within, or at the end of a time interval t_b, when t_b = 1 second. or continuously, within, or at the end of a larger time interval.

And the end of these "instantaneous implosions" are happening (to this same limited inertial velocity displaced mass +m). Again, at the "instantaneous discontinuance" of this same (classical) limited force: +F_v. Instantaneously, within, or at the end of a time interval t_b, when t_b = 1 second. Or continuously, within, or a the end of a larger time interval.

And since, now if there is a "velocity-momentum" displacement "of increase", for a limited inertial velocity displaced mass +m—this implies a "mass-momentum" displacement "of decrease"—for this same limited inertial velocity displaced mass +m.

And also, if there is a "volume-momentum" displacement "of increase", for a limited inertial velocity displaced mass +m—this implies a "density-momentum" displacement "of decrease"—for this same limited inertial velocity displaced mass +m.

And we can now show the other new kinematic momentum relations between the four new kinematic momentum measuring formulas: (20), (21), (22), and (23).

And which we can put: is that a limited inertial velocity displaced mass +m is experiencing an "instantaneous increase" in its inertial *velocity-momentum displacement*. And then, this same limited inertial velocity displaced mass +m is now experiencing an "instantaneous decrease" in its inertial *mass-momentum displacement*. As a result of the same limited inertial accelerating displaced mass +m. Instantaneously, at the end, or within, a time interval t_b, when t_b = 1 second. Or continuously, within, or at the end of a larger time interval.

And which the end of these two empirical conditions—(for the *now* limited inertial velocity displaced mass +m): are also due, or caused by the end of the instantaneous discontinuance of this same (classical) limited force: +F_v. Instantaneously, at the end, or within, a time interval t_b, when t_b = 1 second. Or continuously, within, or at the end of a larger time interval. (As this section displays).

I will now offer a new limited kinematic "velocity" measuring formula, which will measure separately the "velocity" aspect of the "velocity-momentum" displacement of formula (20). And I will also offer a new limited kinematic "mass" measuring formula which measures separately the "mass" aspect of the "mass-momentum" displacement of formula (21).

And I will also offer the new limited kinematic "volume" measuring formula which will measure separately the "volume" aspect of the "volume-momentum" displacement of formula (22). And I will also offer the new limited kinematic "density" measuring formula which measures separately the "density" aspect of the "density-momentum" displacement of formula (23).

The instantaneous limited inertial "velocity v" displacement (of decrease/of increase), for a limited inertial velocity displaced mass $+\underline{m}$, is measured and given by the new measuring formula: $(+v)/(+1-(\underline{u}/\underline{c})^2)^{+\frac{1}{2}}$. And where the uniform inertial velocity displaced mass $+\underline{m}$, is cancelled out of formula (20).

And the instantaneous limited inertial "mass $+\underline{m}$" displacement (of increase/of decrease), for a limited inertial velocity displaced mass $+\underline{m}$ is measured and given by the new measuring formula: $(+\underline{m}v)(+1-(\underline{u}/\underline{c})^2)^{+\frac{1}{2}}$. And where the uniform limited inertial vector velocity term: $+\underline{v}$ is cancelled out of formula (21).

And the instantaneous limited inertial velocity "volume $+\underline{V}$" displacement (of decrease/of increase), for a limited inertial velocity displaced mass $+\underline{m}$—is measured and given by the new measuring formula: $(+\underline{V})/(+1-(\underline{u}/\underline{c})^2)^{+\frac{1}{2}}$. And where the uniform inertial vector velocity term: v, and, and the uniform scalar density term: $+\underline{d}$, as both of these terms are cancelled out of formula (20).

And the instantaneous limited inertial velocity "density term: $\underline{d}$" displacement (of increase/of decrease), for a limited inertial velocity displaced mass $+\underline{m}$, is measured and given, by the new measuring formula: $(+\underline{d})(+1-(\underline{u}/\underline{c})^2)^{+\frac{1}{2}}$.

And both of these two new physical events are due, or are caused by the end of an "instantaneous explosions" happening to this same limited inertial velocity displaced *mass* +m *outwards from itself*. Which happens, instantaneously, within, or at the end of a time interval t_b, when $t_b = (0 \text{ to } 1)$, of one second. Or continuously, within, or at the end of a larger time interval.

And again, the end of these "instantaneous explosions" are happening to this same limited inertial velocity displaced mass +m—at the instantaneous discontinuance of this same (classical) limited force: $+F_v$. Instantaneously, within, or at the end of a time interval t_b, when $t_b = (0 \text{ to } 1)$, of one second. Or continuously, within, or at the end of a larger time interval.

And whereby, when a limited inertial velocity displaced mass +m is experiencing an "instantaneous decrease" in its inertial *mass-momentum displacement*. As a result of the same limited inertial velocity displaced mass +m is experiencing an "instantaneous decrease" in its *density-momentum displacement*.

As both, (as we have seen), of these two new above physical events are due, or are caused by the end of an "instantaneous explosions" of this same limited inertial velocity displaced *mass* +m *outwards from itself*. Instantaneously, within, or at the end of a time interval t_b, when $t_b = (0 \text{ to } 1)$, of one second. Or continuously, within, or at the end of a larger time interval.

And these two empirical conditions of: "instantaneous implosions/ explosions" fully, and completely, describes and explains what will happen to a limited inertial velocity displaced mass +m.

And which the beginning of these two empirical conditions of: "instantaneous implosions/instantaneous explosions" were caused by the *former dynamic actions* of a (classical) limited force: $+F_v$ "instantaneously impacting", or "impacted" upon a limited non-inertial accelerating displaced *volume-force*, $+V(+\underline{\dot{v}})/(+1-(\underline{\dot{u}}/\underline{c})^2)^{+\frac{1}{2}}$, with respect to the uniform scalar volume term: $+\underline{V}$, as both of these terms are cancelled out of formula: (21).

And I also want to note, that the same kind of empirical analysis and physical interpretations I had previously given to the four new limited momentum measuring formulas, (20), (21), (22), and (23)—also similarity applies to these above four new and limited kinematic measuring formulas.

And now according to the four new above limited kinematic $+P_v$ momentum measuring formulas: (20), (21), (22), and (23): and which these formulas measures the end of the extreme "instantaneous implosion effects"—of a real limited inertial *velocity displaced mass +m into itself, which it will experience*. And which at the end of these "effects" are due, or are caused by the instantaneous discontinuance of the intense dynamic actions of a (classical) limited force: $+F_v$—which previously had "impacted" upon this same mass $+m$. And these events are happening, instantaneously, within, or at the end of a time interval t_b, when $t_b = 1$ second. Or continuously, within, or at the end of a larger time interval.

And also according to the four new limited kinematic $+P_v$ momentum measuring formulas: (20), (21), (22), and (23). And which these formulas also measures the end of the extreme "instantaneous explosion effects" of a real limited *velocity displaced mass +m outwards from itself*. And which it will experience. And which the end of these "effects" are due, or are caused by the instantaneous discontinuance of the intense dynamic actions of a (classical) limited force: $+F_v$ Instantaneously, within, or at the end of a time interval t_b, when $t_b = 1$ second.—which previously had "impacted" upon this same mass $+m$. Or continuously, within, or at the end of a larger time interval.

And now, the problem is in regards to the instantaneous end of which one of these two empirical conditions of: "instantaneous implosion effects". And, or of: "instantaneous explosion effects". Which will occur simultaneously, for this real and limited inertial velocity displaced mass $+m$. And this scenario depends upon the *former force-magnitudes* of this same (classical) limited force: $+F_v$. And also depends upon the kind of mass used.

And we must note, that the end of these unique "dynamic action effects" that this same inertial velocity displaced mass $+m$ will experience—are now happening, instantaneously, within, or at the end of a time interval t_b, when $t_b = 1$ second. Or continuously, within, or at the end of a larger time interval.

And we must also note, that the end of these two new empirical conditions of: "instantaneous implosion effects"; and of: "instantaneous explosion effects". (As both of conditions were formerly caused by a limited force). Which must then result in extreme: "thermo kinematic and dynamic effects"—that this same mass $+\underline{m}$ will experience.

As we have seen in the preceding section of this book, involving: "thermo kinematic and dynamic effects"; "intense thermo radiation effects"; "hot plasmatic effects"; and "hot fragments of mass": that a mass $+\underline{m}$ will experience. As caused by the dynamic actions of a limited and real force: $+F_v$, instantaneously impacting, or continuously impacting upon this same mass $+\underline{m}$, within, or at the end of a time interval $\underline{t}_b$, when (0 to 1), of one second. Or continuously, within, or at the end of a larger time interval. These "effects" will happen to a mass $+\underline{m}$ at the end of the: "intense implosion effects"—that the mass $+\underline{m}$ will experience. And the other "effects" that the same mass $+\underline{m}$ will experience are: "intense explosion effects"—that a mass $+\underline{m}$ will "explode outwards from itself". And these physical events are happening, instantaneously, within, or at the end of a time interval $\underline{t}_b$, when $\underline{t}_b = (0 \text{ to } 1)$, of one second. Or continuously, within, or at the end of a larger time interval.

And I must note (as I have shown in the preceding section of this book)—that the event of: vector velocity limit: $+c$; and the event of: scalar speed limit: $+\underline{c}$ of light, in free vacuum. Hence, these terms are with respect to a "converted displaced mass $+\underline{m}$" would result into "intense momentum effects"; and into "intense thermo kinematic and dynamic effects". And these facts are the reason why we have set: $+\underline{m}c = 0$, and $+\underline{mc} = 0$.

And these two above (classical) limit lower-momentum displacement formulas must be the case, and when they were "formerly" caused by the instantaneously dynamic actions of a (classical) limited force: $+F_v$.

And since, when this same (classical) limited force: $+F_v$ "had instantaneously impacted" upon a limited accelerating displaced mass $+\underline{m}$—which is the same mass term $+\underline{m}$ of the two above limit lower-momentum formulas. And this "former" real empirical event of: "instantaneously impacting" is happening, instantaneously, within, or at the end of a new time interval (0 to 1), of one second. Or continuously, within, or at the end of a larger time interval.

And then, when at the end of this time interval t_b, when $t_b = (0 \text{ to } 1)$, of one second. Or continuously, within, or at the end of a larger time interval.

That is, at the end of the instantaneous discontinuance of this same (classical) limited force: $+F_v$—which previously had "impacted" upon this same real mass $+\underline{m}$. And then, the former uniformly accelerated displaced mass $+\underline{m}$, (or whats left of it), has now "instantaneously inertially slides" into being a limit $+c$ vector velocity displaced mass $+\underline{m}$. Or into being a limit $+\underline{c}$ inertial speed displaced mass $+\underline{m}$. That is, when: $(+\underline{mc} = 0)$, and even: $(+\underline{mc} = 0)$.

And these empirical facts tells us that a real limited inertial velocity displaced mass $+\underline{m}$: or a limited inertial speed displaced mass $+\underline{m}$—has been, totally, converted into intense kinematic momentum effects. And also totally converted into incredible intense thermo kinematic effects. (As were formerly caused by the dynamic actions of a (classical) limited force: $+F_v$.

And now, Albert Einstein's special relativistic mass increase measuring formula: $(+\underline{m}_o)/(+1-(\underline{u}/\underline{c})^2)^{+\frac{1}{2}} = +\underline{m}$. And which purports to measure the infinite mass increase, ie., $(+\underline{m} \to \infty)$. As will be experienced by a limited inertial velocity displaced mass $+\underline{m}$, when the limiting inertial factor $(+1-(\underline{u}/\underline{c})^2)^{+\frac{1}{2}}$ tends towards a zero value (0).

And Albert Einstein's special relativistic momentum measuring formula: $(+\underline{m}_o v)/(+1-(\underline{u}/\underline{c})^2)^{+\frac{1}{2}} = +\underline{m}v$, also purports to measure the infinite momentum increase, ie., $(+\underline{m}v \to \infty)$. As will be experienced by a limited inertial velocity displaced mass $+\underline{m}$, when the limiting inertial factor: $(+1-(\underline{u}/\underline{c})^2)^{+\frac{1}{2}}$, tends towards a zero value (0).

And these empirical conclusions must be the case, as based directly upon my interpretations of Albert Einstein's relativistic mass increase measuring formula, as we have seen.

But I have rejected Albert Einstein's special relativistic mass increase measuring formula. And then, I have also rejected his special relativistic momentum measuring formula, as both formulas are false measuring formulas, according to my new mechanical theory.

And nor have I used an inertial rest mass term: $+\underline{m}_o$, (in direct contrast to Albert Einstein's special relativity theory)—in any of the four new limited kinematic $+P_v$ momentum measuring formulas: (20), (21), (22), and (23). And nor in any of the four new limit kinematic $+P_c$ lower-momentum measuring formulas: (24), (25), (26), and (27). (See the end of this section about this point). And the empirical reason why I have not used an inertial rest mass term: $+\underline{m}_o$ in any of my kinematic momentum measuring formulas in this section—or in this book—is based directly upon my use and acceptance of a diagram (8). (Please see the introduction of this book about this matter).

In fact, as we now seen, Albert Einstein's special relativistic momentum measuring formula: $(+\underline{m}_o v)/(+1-(\underline{u}/\underline{c})^2)^{+\frac{1}{2}} = +\underline{m}v$, (with respect to the limiting inertial factor: $(+1-(\underline{u}/\underline{c})^2)^{+\frac{1}{2}}$—is a false momentum measuring formula.

Since, when this limiting inertial factor is numerically analyzed and physically interpreted—and then Albert Einstein's special relativistic momentum measuring formula will give a momentum displacement. That is, when this same limiting inertial factor *is greater than a term*: $+\underline{k}$, and which tends towards a value $\underline{3}$. (Having the unit of measurement of 10^4 kilometers per one second), in the inequality (9)), (see introduction).

And then, Albert Einstein's special relativistic momentum measuring formula will give a momentum increase displacement—which tends

towards an infinite momentum increase displacement. That is, when this same limiting inertial factor: *is less than a term*: $+\underline{k}$—and which tends towards a zero value (0), (having the unit of measurement of 10^4 kilometers per one second), in the inequality (9), (see introduction).

And it is because of these facts, there is no real empirical interpretations I can give to Albert Einstein's special relativistic momentum measuring formula. This is because, empirically, it is a false momentum measuring formula—according to my new mechanical theory.

But these two special relativistic conclusions, (about his special relativistic mass increase measuring formula; and his special relativistic momentum measuring formula). Whereby, as it is practiced by Albert Einstein, and by the many proponents of his special relativity theory, it is in actuality: "outrageous physics". But this "outrageous physics" of a real limit, or of a real limited velocity displaced mass +$\underline{m}$, "increasing towards infinite mass", as it is based or determined by a limiting inertial factor: $(+1-(\underline{u}/\underline{c})^2)^{+\frac{1}{2}}$. And where this inertial limiting factor tends towards a zero (0) value *is extremely far from the truth of this matter*. And this special relativistic mass increasing measuring formula, does in actuality, leads to the blatant violations of our basic laws of physics.

Since, we can ask: where does this infinite increase of inertial mass come from? It must only be inertially created by all real and limited mass; or by all real and limited inertial velocity displaced cosmic mass: having their own inertial velocity displacement. Which is ridiculous. And when this question is thought about, one becomes aware that his special relativistic mass increase measuring formula, does in actuality, contradict the mass-conservation laws of physics. Whereby, mass-energy is neither created nor destroyed.

But it is this kind of anomalistic thinking and reasoning on Albert Einstein's part, and on the part of his many proponents of his special relativity theory, which had lead them to this "outrages physics" of a limit, or a real and limited inertial velocity displaced mass +$\underline{m}$ increasing towards infinite mass increase, ie., (+$\underline{m}$ → ∞). But I have rejected all the mechanical results (all the dynamic and kinematic results), of Albert Einstein's original special relativity theory. And thus, I reject, totally, all of his special relativistic measuring formulas, as being false measuring formulas, with respect to my new non-relativistic mechanical theory, I have proposed and advanced in this book.

In concluding this section, as I have offered the four new limited kinematic +P_v momentum measuring formulas (20), (21), (22), and (23), I will now offer the four new limit kinematic +P_c lower-momentum measuring formulas. And we now have, (in regards to a mass term: +$\underline{m}$):

$$(+\underline{m}\underline{c})/(+1-(\underline{c}/\underline{c})^2)^{+\frac{1}{2}} = +P_{c}' = 0) \tag{24}$$

And:

$$(+\underline{m}\underline{c})(+1-(\underline{c}/\underline{c})^2)^{+\frac{1}{2}} = +'P_{c} = 0) \tag{25}$$

And now we have, (in regards to a volume term: $+\underline{V}$, and of a density term: $+\underline{d}$, with respect to a mass term: $+\underline{m}$), the other two new limit kinematic lower-momentum formulas:

$$(+\underline{V}\underline{c})/(+1-(\underline{c}/\underline{c})^2)^{+\frac{1}{2}} = (+P_{c}^{*} = 0) \tag{26}$$

And:

$$(+\underline{d}\underline{c})(+1-(\underline{c}/\underline{c})^2)^{+\frac{1}{2}} = (+^{*}P_{c} = 0) \tag{27}$$

And where $|+c| = +\underline{c}$. And where the term: $+\underline{V}$ is scalar volume, and where the term $+\underline{d}$ is scalar density. And where by the formula: $(+\underline{V})(+\underline{d}) = +\underline{m}$, a scalar mass term.

And where the new limit kinematic $+P_{c}'$ lower-momentum formula (24) measures the *instantaneous limit-velocity-momentum displacement* (of zero decrease/of zero increase), of a mathematical non-existent displaced mass $+\underline{m}$, of a time interval of one zero second.

And where the new limit kinematic $+'P_{c}$ lower-momentum formula (25) measures the *instantaneous limit mass-momentum displacement* (of zero increase/of zero decrease), of a mathematical non-existent displaced mass $+\underline{m}$, of a time interval of one zero second.

And where the new limit kinematic $+P_{c}^{*}$, lower-momentum formula (26) measures the *instantaneous volume-momentum displacement* (of zero decrease/ of zero increase), of a mathematical non-existent displaced mass $+\underline{m}$, of a time interval of one zero second.

And where the new limit kinematic $+*P_c$ lower-momentum formula (27) measures the *instantaneous density-momentum displacement*, (of zero increase/of zero decrease), of a mathematical non-existent displaced mass $+\underline{m}$, of a time interval of one zero second.

And it is because of the limiting inertial factor: $((+1-(\underline{c}/\underline{c})^2)^{+\frac{1}{2}} = 0)$, for these four new limit kinematic $+P_c$ lower-momentum measuring formulas: (24), (25), (26), and (27). And then there is no real empirical interpretations I can give to these four new limit kinematic lower-momentum measuring formulas: but only a mathematical interpretations.

And to preserve the form of this book, is the only reason I have included these four new limit kinematic $+P_c$ lower-momentum formulas: (24), (25), (26), and (27), in the text of this book.

But I should note, that these same four new limit kinematic $+P_c$ lower-momentum measuring formulas: (24), (25), (26), and (27), serves as a natural lower-limits for the four new "limited" kinematic momentum measuring formulas: (20), (21), (22), and (23). And this point then concludes this section.

CONCLUSION

I have not in this book rejected the special relativistic conceptions of "inertial relative reference frames", and nor of: "inertial relative observers", and thus, not of: "Lorentz-Einsteinian transformation equations". (See appendices about this point). But I have rejected the incomplete use of these same special relativistic conceptions Albert Einstein used in his special relativity theory. Instead of the conception of "relativity", I will use the new conceptions of: "limit" and "limited".

And many physicists use these same incomplete special relativistic conceptions in their physic's research programs. As based upon their total acceptance of Albert Einstein's original special relativity theory. Open any physic's textbook to see my point.[10] But these same physicists consider Albert Einstein's original special relativity theory to be complete and consistent, in regards to their use of these same special relativistic conceptions. And they would argue as such, as based upon their total acceptance and their understanding of Albert Einstein's original special relativity theory. But since, I have rejected Albert Einstein's original special relativity theory to be incomplete and inconsistent, then the burden of proof for this conclusion falls on me.

But I have not only rejected Albert Einstein's original special relativity theory to be incomplete because of its use of these same special relativistic conceptions. But I have rejected all the mechanical results, (all the kinematic and dynamic results), of Albert Einstein's original special relativity theory. With respect to my new non-relativistic mechanical theory: which I have proposed and advanced in this book.

All of Albert Einstein's special relativistic measuring formulas are false measuring formulas, according to my new non-relativistic mechanical theory. And I have offered in this book, new mechanical measuring formulas, and I will offer, in the appendices, new energy measuring formulas, and so on. To replace all of Albert Einstein's original special relativistic measuring formulas.

And then, I will, also, in the appendices of this book, propose and advance a new theory, which will use such empirical conceptions of "limit", and "limited" inertial, and non-inertial reference frames. And also offer four new kinds of "limit" and "limited" non-inertial observers. And also offer four new kinds of "limit", and "limited" transformation equations.

In these appendices I will use the new and revised Newtonian first law of motion (4). As well as of the two new empirical conceptions of "limited forces", and of the instantaneous discontinuance of these same "limited forces". And perhaps, also of "limit forces", and of the instantaneous discontinuance of these same "limit forces".

And this new theory will be based directly upon the principles and results of my new mechanical theory I have proposed and advanced in this book. And this new theory will be more complete, and true, and empirically sound, than Albert Einstein's original special relativity theory. Which is an incomplete and false scientific theory.

And yet, the question needs to be asked: why have I proposed and advanced in this book a new non-relativistic mechanical theory which does not use any relativistic conceptions?

My answer to this question is: that my new non-relativistic mechanical theory, which I have proposed and advanced in this book, is more empirically true, consistent, and complete, than Albert Einstein's special relativity theory. My second answer is: I question the relativistic conceptions of: "Lorentz-Einsteinian transformation equations". As these two conceptions may not have any sound basis in Nature. And in the appendices of this book, we will fully examine these relativistic conceptions as to whether they have any sound empirical basis in Nature. Or whether the relativistic conceptions of: "relative inertial reference frames", are only a mathematical concept, and does not have a sound empirical basis in Nature.

And since, I have rejected Albert Einstein's special relativity theory in this book, hence, Herman Minkowski's original 4th dimensional space-time theory is also rejected: as being empirically incomplete, and inconsistent.

This is because, Herman Minkowski's original 4th dimensional space-time theory is based directly upon Albert Einstein's original special relativity theory,[11] which I have completely rejected all of his special relativistic dynamic and kinematic results in this book. And I have proposed and advanced a new non-relativistic mechanical theory to replace Albert Einstein's original special relativity theory.

Although, this conclusion I make about Herman Minkowski's original 4th dimensional space-time theory as being empirically incomplete and inconsistent. This conclusion I make is very controversial on my part. This is because Herman Minkowski's 4th dimensional space-time is very well established in all areas of physics. And it is completely accepted by most, if not by all physicists.[12] And also, Herman Minkowski's original space-time theory is used in an extended manner, in Albert Einstein's original General Relativity Theory, which is an all encompassing theory of gravitation.[13] And it may be that we must generalize Herman Minkowski's original 4th dimensional space-time theory, as being based upon this new theory, or model, I will propose and advance in the appendices of this book.

And I can note, that in Albert Einstein's original special relativity theory for all inertial motive relative observers. And for all inertial motive relative reference frames are to be measured, with respect to other inertial motive relative reference frames. And also with respect to other inertial motive relative observers.

But according to my new mechanical theory I have proposed and advanced in this book, all "limit", and "limited" inertial, and non-inertial motive displaced cosmic masses: are to be measured with respect to the uniform inertial scalar constant speed limit $+\underline{c}$ of light, in free vacuum. And this speed limit $+\underline{c}$ of light, in free vacuum is: a fundamental constant of Nature, and also a upper speed[14] limit for all limited motive displaced cosmic masses, in free vacuum.

And also, according to my new mechanical theory, all limited (inertial, and non-inertial) motive displaced masses, are to be measured—with respect to the uniform limit dynamic lower-force formula: $(+F_c = 0)$. And also with respect to the uniform limit kinematic lower-momentum formula: $(+P_c = +\underline{mc} = 0)$, or: $(+\underline{P}_c = +\underline{mc} = 0)$ And these two limit measuring formulas, are the fundamental and constant *lower-motive limits for all limited* (inertial, and non-inertial) motive displaced cosmic masses.

In conclusion, in the appendices of this book, I will offer new mass-energy measuring formulas to replace the rejected special relativistic energy measuring formulas. And also in this appendices, I will give, a complete and empirical resolution of the "twin paradox", or of the "clock paradox", which were derived from the special relativity theory.

REFERENCES AND NOTES

[1] Emilio Serge, *From X-Ray to Quarks*, (University of California Press, San Francisco, California, 1980), page 80.

[2] Isaac Newton, *Mathematical Principles of Natural Philosophy*, (Robert Maynard Hutchins, Ed., (The Great Books of the Western World), bk., 34)), (University of Chicago Press, Chicago, Illinois, 1984), pages 1-369.

[3] Albert Einstein, *On the Electrodynamics of Moving Bodies*, (W. Perrett, and G.B. Jeffery, Ed's., "The Principle of Relativity" (Dover Publications, Inc., New York, N. Y., 1982), pages 141-142.

[4] Abraham Pais, *Subtle is the Lord . . . the Science and Life of Albert Einstein*, (Oxford University Press, New York, N. Y., 1983), pages 141-142.

[5] Robert Resnick and David Halliday, *Basic Concepts in Relativity and Early Quantum Theory*, (Macmillan Publishing Company, New York, N. Y., 1993), pages 115-116.

[6] Thorton Page, *Newton's Laws*, (Rita G. Lerner, and George L. Trigg, Ed's., Encyclopedia of Physics), (Addison-Wesley Publishing Company, Inc., Reading, Massachusetts, 1981), page 661.

[7] Giles Cohen-Tannoudji, *Universal Constants in Physics*, (translated by Patricia Thickstun), (McGraw-Hill, Inc., New York, N. Y. 1993), pages 7-36.

[8] Claude Kaeser, *Relativity, Special Theory*, (Rita G. Lerner, and George L. Trigg, Ed's., Encyclopedia of Physics) (Addison-Wesley Publishing Company, Inc., Reading, Massachusetts, 1989), pages 886.

[9] The following formula: $(+\underline{m})(+1-(\underline{u}/\underline{c})^2)^{+\frac{1}{2}}$ is for "mass displacement" (of increase/of decrease). And while the following formula: $(+\underline{v})/(+1-(\underline{u}/\underline{c})^2)^{+\frac{1}{2}}$ is for "velocity displacement" (of decrease/of increase).

And we arrive at these two formulas by the following formulas: $(+\underline{mv}/+\underline{v})(+1-(\underline{u}/\underline{c})^2)^{+\frac{1}{2}}$ and: $(+\underline{mv}/+\underline{m})/(+1-(\underline{u}/\underline{c})^2)^{+\frac{1}{2}}$.

9   The following formula: $(+\underline{d})(+1-(\underline{u}/\underline{c})^2)^{+\frac{1}{2}}$ is for "density displace-
    ment" (of increase/of decrease). And while the following formula:
    $(+\underline{V})/(+1-(\underline{u}/\underline{c})^2)^{+\frac{1}{2}}$ is for "volume displacement" (of decrease/of
    increase).

    And we arrive at these two formulas by the following formulas:
    $(+\underline{m}/+\underline{V})(+1-(\underline{u}/\underline{c})^2)^{+\frac{1}{2}}$ and $(\underline{m}/+\underline{d})/(+1-(\underline{u}/\underline{c})^2)^{+\frac{1}{2}}$.

10 Richard Philips Feynman, *The Feynman Lectures on Physics*, (Addison-Wesley
    Publishing Company, Inc., Reading, Massachusetts, 1989), Volume I: 15-1
    to 15-t; 16-1 to 16-5; 17-1 17-5.

11 J. L. Synge, *Relativity, the Special Theory*, (North-Holland Publishing
    Company, Amsterdam, Holland, 1965) pages 1-36.

12 Edward R. Harrison, *Cosmology: the Science of the Universe*, (Cambridge
    University Press, New York, N. Y., 1981) pages 123-159.

13 Joshua N. Goldberg, *Space-Time*, (Rita G. Lerner and George L. Trigg, Ed's.,
    Encyclopedia of Physics, 1981), 946-948

14 The speed limit $+\underline{c}$ of light, can be a fundamental speed limit for all inertial
    motion displaced cosmic masses, which has their inertial motions, by the
    actions of forces.

    Or the speed limit $+\underline{c}$ of light can serve as a fundamental lower-speed
    limit—when the term $+\underline{c}$ is multiplied into a mass term: $+\underline{m}$, where then,
    $+\underline{mc} = 0$.

APPENDICES

APPENDIX A

THE NON-EXISTENT UNIVERSE: FOR FIRST HAND HUMAN EMPIRICAL EXPERIENCES

-a-

As we have biological existence, we then have living first hand empirical experiences of our earth, and of our solar system. We empirically experience, at first hand, the light and heat of our sun. Without the sun we would not have human existence. And because of our existence, we then have first hand human empirical experiences of our earth, and of our sun. The sun warms and gives us light, in which we use to empirically experience, or perceive, at first hand, our earth, and of our sun, as well as comets, and of our planets of our solar system.

We also perceive at night, the "great electro-magnetic radiations universe", which is the greatest "electro-magnetic radiations universe show" for all time: "the great light universe!". But this is all we do experience, empirically, about our universe, only the "light universe", and no more. We don't empirically perceive with our eyes the higher electro-magnetic radiations frequencies of our universe.

Because of the great spatial and temporal distances between the galaxies, and between the stars, of our universe—we are then unable to empirical experience, at first hand, to these many galaxies and stars. But we can, empirically, experience, at first hand, our planetary systems as astronauts—by traveling within our solar system by using rocket engine propulsion space-crafts.

Travel in our universe is not physically feasible which is due to the great spatial and temporal distances of the stars and galaxies.[1]

Please see Appendix C: "the empirical resolution of the twin paradox". Which says that no rocket propelled space-crafts, or any kind of space-crafts, can never come close to the uniform speed limit c of light, as it is measured by one second, in free vacuum space and time. This is because, not that the mass of the space-crafts would approach "infinite mass", but because the self-propelled space-crafts, having high accelerations, would destroy the observers of the space-crafts, and also destroying the space-crafts, into becoming hot burning fragments of mass and energy.

And neither does time slow down (dilate) when a time piece is highly accelerated in a space-craft, but instead the time piece is destroyed into hot burning fragments of mass and energy. Which is due to the intense actions of the space-craft's rocket engine.

Our only possibility for space travel would be robots, having artificial intelligence, and where the robots would repair themselves and their space-crafts, and also repairing their rocket engine. One way or another, people will also seek ways to travel in space and time. We are explorers, and will find ways and means to travel in small parts of our universe. Such as by traveling within our Milky Way Galaxy.

I can hear the fundamentalists saying: "Dennis told us that our universe does not exist". And thus, the question needs to be asked: "why do we exist?"[2]

-b-

To be more serious, and less flippant, this thesis I am proposing implies that we have a new third "Copernicus Revolution", which is: "the non-existent universe: for first hand human empirical experiences". The first "Copernicus Revolution" when Copernicus proposed an empirical thesis that instead of the Earth being at the center of the solar system, the sun was at the center of the solar system. And the planets revolved around the sun. The second "Copernicus Revolutions" is when Charles Darwin successfully argued his thesis, with many empirical examples, that all life on the Earth, had evolved from simpler life creatures.

The Darwinian revolution challenged the Judaeo-Christian's beliefs that a supreme and omnipotent God created our Earth, our sun, and planets, our entire universe, and all the life on Earth, including Human Life, Fundamentalists call this thesis "intelligent design", or "creationism".

When astronomers observe the night sky with their telescopes, and take "photo-graphs" of what they observe. The question is then asked: are these "photo-graphs", when we view them, or scan them, the "real things" of our universe? These "photo-graphs" of parts of our universe seems real and close to us, and where these "photo-graphs", and "pictures" seem to us, to be the "real McCoys". Yet we view and scan figurative paintings and drawings also seem real and alive, but they are not so. When we view the universe with our telescopes, we see many galaxies, stars, nebula, and other strange phenomena of our universe. But these "images", the "photo-graphs", of our electro-magnetic radiations universe are very small in scale. And yet, we enlarge our electro-magnetic radiations universe, or parts of it, by the use of our telescopes, and by taking "photo-graphs" of these parts. When we view parts of our universe with our naked eyes, the stars seem very small in scale, and very distant in space and time. What we mostly observe with our naked eyes is the vastness of space and time, and the vastness of distance, with respect to our universe. But even with our enlargement of the stars, galaxies, nebula, and other strange phenomena of our universe: by the use of our telescopes, and radio telescopes, we still do not experience, at first hand, our empirical universe.

Perhaps, we can conclude, when we view parts of our universe, the stars, galaxies, nebula, and other strange phenomena of our universe, by the use of our telescopes and radio telescopes. That is, we are only viewing parts of our universe, at a second hand human empirical experiences. Or perhaps, a third hand human empirical experiences. And thus, perhaps, only an *nth* hand human empirical experiences.

A photo-graph is a image in color, or in black and white, by the use of light radiations, from parts of our universe. These photo-graphs are very similar to a drawing made in colored ink, onto a piece of paper.

The photo-graphs of sections of our universe is similar to an astronomer viewing through a telescope, and at the same time making colored drawings, or making black and white drawings, of what he sees in these parts, or sections, of our universe.

-c-

Nevertheless, with all of our speculations about our "astronomer", we can never experience, as life, at first hand, in an empirical sense, our total universe, or even small sections of it. The universe simply does not exist for our first hand empirical experiences of it. What does exist for us, is our solar system, which includes our sun, planets, comets, our earth, our space and time, with respect to our solar system.

The most we can do, in future times, is to send intelligent robots, within special kinds of space-crafts, to visit some of the stars of our Milky Way Galaxy. But it would take hundreds, or millions of light years for the robotic space-crafts to accomplish their missions.

It might be argued, that as we have first hand experience, in a human sense, of our Earth, of our solar system, that we then can *extrapolate* that most of the stars of our Milky Way Galaxy are similar with our sun, and similar with our solar system. Hence, we could conclude: that there may be life on the planets of these other many stars.

But our method of "extrapolation" is not of an "empirical first hand human experiences" for these other solar systems of our Milky Way Galaxy. Our method of "extrapolation" is only a "conjecture", a "guess", a "inductive inference", and it is nothing "certain", nothing "necessary", nothing "final". The universe is a extreme vast expansion in space and time—which we do not experience, empirically, at first hand. It may be that some of the parts of our universe may be "non-existent" for first hand human experiences. And for other parts of our universe, or of our Milky Way Galaxy, may be "existent" for kinds of life, other than our own, in regards to other suns, and their solar systems. But in regards to these special parts, of our milky way galaxy, they may have their own life, their own sun, and their own solar system, and yet, we cannot travel to.[3] This is because, according to Appendix C, the space and time travel, in our universe, by various kinds of propulsion engines: such as by force rocket engines is not empirically feasible.

Notes

1   "The more the universe seems comprehensible, the more it seems pointless". Steven Weinberg, who made this statement in his book: *The First Three Minutes*, p. 154. Published by Basic Books, New York, N. Y.

   The universe by itself, and for itself, in regards to the natural forces of Nature, and of the material and electro-magnetic sub universe—is meaningful, even though we are unable to experience the universe by first hand human empirical experiences. Some people, if not all people, that this thesis, or this model of the universe, which I have proposed, may make many people feel very much alone when they view the night sky. But it is better to have the truth, then not have the truth, about our place in this world we call Earth, or our place in the entire universe, which is the "great light universe, we experience every night.

   As intelligent people, we can view the great night sky as the "great light universe", and thus, we experience the "great now time", with respect to the great light universe, with respect to the speed limit +$\underline{c}$, of light, throughout the total universe. And these facts are not "pointless", and nor "meaningless".

2   The physicist Amit Goswami, in his book: *The Self-Aware Universe*: *How Consciousness Creates the Material World*. Published by Jeremy P. Tarcher/Putnam, New York, N. Y. I cannot agree with the title of Doctor Goswamin's book. Since, if there were no intelligent life in the universe, the universe would still continue to exist. And the universe would still follow all the natural laws of Nature in its existence. The material world, the universe, existed a long cosmic time, before time, ie., before there were intelligent life, to perceive it, or observe it.

3   Many physicists in today's time and age, use the empirical and mathematical concepts such as "ten dimensions", "super strings", "spacetimes", "wormholes", "blackholes", "whiteholes". As an example of these kinds of concepts, see Brian Green's book: *The Elegant Universe*, by W. W. Norton & Company, 500 fifth street.

APPENDIX B

THE PRINCIPLE OF CONTINUITY

-a-

Centuries ago, Galileo performed an experiment of carrying two unequal masses up the stairs of the leaning tower of Pisa. And at the top of the tower, he dropped the two unequal masses at the same height, and at the same time. And he saw that the two unequal masses fell side by side, and both masses hit the ground simultaneously.

Albert Einstein centuries later had argued from Galileo's experiment to put forward a "theory of equivalence" to explain Galileo's experiment. But instead of formulating a new theory to explain Galileo's experiment, as Einstein had done, I will instead explain, empirically, Galileo's experiment by the use of applied forces, and limited masses, within a gravitational field.

When two unequal bodies in mass, where $+\underline{m}_1$, and $+\underline{m}_2$, and where $+\underline{m}_1 > +\underline{m}_2$. And where we have two limited and applied unequal forces: $+F_1$, and $+F_2$, where: $+F_1 > +F_2$. Hence, when these same two unequal forces—the above two unequal forces—are acted upon these two unequal masses, the above two unequal masses. And thus, whereas, when the force: $+F_1$ is acted upon the mass: $+\underline{m}_1$. And at the same time, when the force: $+F_2$ is acted upon the mass: $+\underline{m}_2$. And thus, both masses $+\underline{m}_1$ and $+\underline{m}_2$ will achieve the exact same height within a gravitational field. And lo and behold, both of these two unequal masses, will hit the ground of the Earth as the same exact simultaneous time. And this empirical fact can be called the "principle of continuity" between two unequal masses, and between two unequal forces.

And thus, this fact can be explained as being two unequal motions, of the two unequal masses, as a result of having two unequal forces, with respect to the Earth's gravitational field.

Hence, given the two unequal masses, $+m_1$, and $+m_2$, but having two equal forces: $+F_1$ and $+F_2$, where: $+F_1 = +F_2$. And whereas, these two unequal masses will achieve two unequal heights, $+h_1$, and $+h_2$, where $+h_1 \neq +h_2$—within the Earth's gravitational field. And thus, when these two unequal masses begin to fall, within the Earth's gravitational field, *they will not hit the ground of the Earth at the same exact simultaneous time.* And this empirical fact can be called the "principle of non-continuity".

-b-

There is a mass $+m_0$ called the "inertial mass", which is to quote: "the mass of a body is a constant which is characteristic for its behaviour under the influence of any force, it is the ratio of force to acceleration". "Likewise, the gravitational force is the product of the 'gravitational field strength' the negative gradient of the gravitational potential, and the mass of the particle. In its role of as a "gravitational charge", we shall call the mass, the "gravitational mass", of the particle. According to Newton's theory of gravitation, the inertial mass, and the gravitational mass of the same body are always equal. This proposition is called "the principle of equivalence . . ."[1]

However, this principle of equivalence is only valid when my principle of continuity is satisfied. Otherwise, it is meaningless and useless to explain Galileo's original experiment. Albert Einstein, when he invented the principle of equivalence, said: "the happiest thought of my life".[2]

Notes

[1] The quote is from: *Introduction of the Theory of Relativity*, by Peter Gabriel Bergmann, pps., 152-153. Published by Dover Publications, Inc., New York, 1976.

I had resolved or explained Galileo's original experiment in 1983, and I realized that perhaps I could make original contributions in physics. But at that time, I wrote an article about it, with lots of graphic notations, and sent it to some journals of physics, it was rejected and returned to me. I put the article away, and went on with other activities.

Although, I did not mention it in the text of this Appendix B, the masses: $+m_1$ and $+m_2$, and the forces: $+F_1$ and $+F_2$, are proportioned to one another, so the masses: $+m_1$ and $+m_2$, will achieve the exact same height within the Earth's gravitational field, and the two masses will fall side by side, simultaneously with the same rate of fall, towards the Earth's ground.

2   The quote is from: *Subtle is the Lord: . . . The Science and Life of Albert Einstein*, by Abraham Pais, pps., 177-179. Published by Oxford University Press, New York, 1992. According to Edward R. Harrison, in his book: *Cosmology: The Science of the Universe*. Pps., 162-164. Published by Cambridge University Press, New York, 1981. Doctor Harrison states that there are two principles of equivalence, the Newtonian and the Einsteinian.

However, it still does not matter whose principles of equivalence is used—since, both principles are useless and meaningless to account for, or to explain Galileo's original experiment.

Many physicists have concluded that "inertial mass" is equivalent to "gravitational mass". The only equivalence between these two kinds of masses is that these travel, or fall, simultaneously, side by side, with the same rate of fall, within the Earth's gravitational fields. Or they travel side by side within a spacecraft's container—in which the spacecraft has a uniform acceleration displacement. Or the spacecraft is traveling with an inertial speed displacement. Or is traveling with an inertial velocity displacement. The two unequal masses will fall side by side. And this fact is called the principle of continuity.

Objects of masses, $+m_1$ and $+m_2$, which have different quantities of mass, and having different forces: $+F_1$ and $+F_2$. And whereas, the two different forces act upon the two different masses, such as: the force: $+F_1$ acts upon the mass $+m_1$, and the force: $+F_2$ acts upon the mass $+m_2$. And thus, both of these two masses will fall side by side, simultaneously, at the same rate of travel.

And when both forces are released, in free vacuum, the two masses will inertially slide into being uniform inertial speed displacements. Or into being uniform inertial velocity displacements. And thus, they will travel side by side, simultaneously, with the same rate of travel, within free vacuum.

In the McGraw-Hill *Encyclopedia of Physics*, second edition,. Sibil P. Parker, Editor in Chief, *The General Theory, Principle of Equivalence*, p. 1199. In the early twentieth century, a physicist, named R. Von Eotvos continuously made experiments, for a decade or more, testing Galileo's original tower of Pisa experiment. Eotvous found that Galileo's experiment was true to a few parts in 10^8. Eotvos tested all kinds of material bodies. Unfortunately, Eotvos wasted his time and effort. Since, as we have seen, Galileo's original experiment can be explained in a simple way, as we have done so in this Appendix B. Instead of supporting Albert Einstein's original *Principle of Equivalence*, we have, instead, put forth a new *Principle of Continuity* to explain, and to account for Galileo's experiment at the top of the tower of Pisa.

APPENDIX C

THE EMPIRICAL RESOLUTION OF THE TWIN PARADOX

-a-

The relativistic identical clock paradox, or the relativistic identical twin paradox: has originated from the special theory of relativity, because of the special principle of relativity—in which the special relativity theory is based upon. And because the special relativity theory implies the conception of: *relative-symmetry* to hold between two identical time events.

And yet, a fundamental paradox has resulted which involves the "relativistic-time-events"; and also involves the idea of "relativistic-reference-frames"—which are considered, theoretically, to approach the uniform vector velocity limit +c, as it is measured by one second. Or to approach the uniform scalar speed limit $+\underline{c}$ of light, as it is measured by one second, in free vacuum space and time.

The identical clock paradox, or the identical twin paradox is as follows: two identical clocks: $\underline{c}_1$, and $\underline{c}_2$ or two identical twins: $\underline{t}_1$ and $\underline{t}_2$—possess the same identical time on one of the relativistic reference frame R'.

And whereas, say both of the twins $\underline{t}_1$ and $\underline{t}_2$; or both the identical clocks $\underline{c}_1$ and $\underline{c}_2$ are on this reference frame R'. And then, if it so happens that one clock $\underline{c}_1$, or one of the twins say $\underline{t}_1$: is instantaneously, yet uniformly accelerated displaced in another relative reference frame 'R (a relative spacecraft). And which this relative reference frame 'R, will, according to the special relativity theory, approach very closely to the uniform scalar speed limit $+\underline{c}$ of light—as caused by the dynamic actions of an applied force: $+F_v$. That is, as caused by the *dynamic force-thrust* of the rocket engine of a relative spacecraft 'R. And this event, happens, instantaneously, within a time interval from (0 to 1), of one second. Or it happens, instantaneously, within a larger time interval.

And thus, there is a relativistic time dilation effects formula for the clock c_1, or for the twin t_1, and which can be given by the kinematic time dilation formula:

$$t = (t_o) / (+1 - (u / c)^2)^{+\frac{1}{2}} \tag{C.1}$$

Where the term: t_o represents the inertial rest time, and where: $+t <$ $+t_o$, and where: $t \to 0$.

And where this relativistic formula (C.1) tells us that the clock c_1, or the twin t_1, has experienced a real time dilation effect, which is the slowing down of the clock c_1 time, or the slowing down of the twin t_1 biological time.

And thus, the identical clock paradox, or the identical twin paradox becomes evident, when we inquire with respect to what other clock, or with respect to what other twin, has the clock c_1, or has the twin t_1 biological time, which has been dilated? And our answer, according to special relativistic conceptions, is with respect to the other clock c_2, or with respect to the other twin t_2 biological time. And which the clock c_2 or the twin t_2 are within another relative reference frame R'. Such as the relative reference frame of the Earth, and which can be considered at relative inertial rest.

And yet, as based upon the special relativistic principles, and of the conception of: *relative-symmetry*, which is implied from the special principle of relativity. And then, it could be concluded that instead of the clock c_1, or the twin t_1, which has experienced the time dilation effects. It was, instead, the clock c_2 time, or the twin biological t_2 time, with respect to their relative reference frame R'—such as, the relative reference frames of the Earth—which has experienced the time dilation effects. *And this is the relativistic twin paradox, or the relativistic clock paradox.*

The paradox has resulted because of the special principle of relativity implies the relativistic conception of: *w-symmetry* to hold and to be empirically valid for all relativistic time events, and which involves the conception of relative-inertial-reference frames, through out free vacuum space and time.

The identical clock paradox, or the identical twin paradox is usually dismissed by proponents of the special relativity theory as being an incorrect handling of the Lorentz-Einsteinian transformation equations. And therefore, it is concluded that there is a relativistic asymmetry involving time dilation effects. And thus, the fallacy of the paradox is revealed. Other resolutions are to appeal to the General Theory of Relativity to resolve the paradox. And still other resolutions make use of the new relativistic conceptions, such as "relative-duration", and so on, to resolve the paradox.

However, the basic point of the identical clock paradox, or of the identical twin paradox, is that the relativistic conception of: *relative-symmetry* cannot be fully applied to the fact of relativistic time dilation effects. And therefore, neither the special principle of relativity, nor the special relativistic conception of: *relative-symmetry* cannot be totally empirically valid in describing, or resolving the many problems in relativistic time events.

And thus, it is our basic arguments that the identical clock paradox, or the identical twin paradox, reveals serious problems in the basic tenets of the special relativity theory. Since, the paradox cannot be fully resolved, in the showing of its fallacy, as based upon the basic tenets of the special relativity theory. In fact, the paradox is a serious "anomaly" of the special relativity theory. The fact is that the identical clock paradox, or the identical twin paradox, has now revealed serious problems with the special relativity theory.[1]

In the following two sections:—b—and—c-, we will offer a kinematical and dynamic resolution of the twin paradox, as well as the clock paradox. And we will also offer a new kinematic and dynamic electro-thermo resolution of these two paradoxes. We intend to show that the fallacy of the identical clock paradox, or of the twin paradox, would never had been formulated: if the special relativity theory had not been formulated and accepted by physicists.

-b-

We are rejecting the relativistic formula (C.1), for the relativistic time dilation formula, as being too imprecise, and being empirically false.

As based upon our previous results of this first book, (concerning whenever an applied force: $+F_v$ is used upon any inertial cosmic rest mass $+m_o$—to instantaneously, yet uniformly accelerate this same cosmic mass $+m_o$, in free vacuum space and time. And which this same mass $+m$ now begins to approach, small, or large, the uniform inertial speed limit $+c$, as it is measured by one second, in free vacuum space and time.

And thus, there would result unique kinds of kinematic and dynamic effects which would occur for this cosmic mass $+m$. (The previous uniform inertial cosmic rest mass $+m_o$). Some of these kinematic and dynamic effects, would be: the instantaneous mass displacements (of increase/ of decrease). As well as instantaneous acceleration displacements of (decrease/of increase), which are happening to this same cosmic mass m. And most important there would be unique kinds of thermo kinematic and dynamic effects happening to this mass—as based upon the tremendous force: $+F_v$ upon the mass m, or upon a relative space-craft 'R.

Hence, it is now easily seen that any relative space-craft 'R, which contains one identical clock, or one identical twin, would have been electro-thermo kinematically and dynamically incinerated into extremely hot and burning mass, and crushed beyond recognition. As based upon a extreme powerful rocket engine, the force: $+F_v$, which is impacting, or had impacted upon this same relative space-craft: 'R.

And thus, the relative identical twin, in this relative space-craft: 'R, *would be dead*. And then it is empirically meaningless to compare any kind of relative time dilation effects, with respect to the living identical twin who has remained safely upon the Earth. And as for the identical clock, in this relative space-craft 'R, would have been crushed into a hot burning mass. And so these new empirical facts resolves the identical twin paradox, and the identical clock paradox.

We see that these paradoxes would never have been formulated if the special relativity theory had taken into account the dynamic actions, or the dynamic empirical actions of a force: $+F_v$ impacting upon these inertial relative reference frames. But in order for the proponents of the special relativity theory to do so, the entire rejections of Albert Einstein's special theory of relativity would have been necessary.

-c-

These facts now brings us to the current interests and their proposals for interstellar travel by means such as anti-matter* matter space-crafts[2], and by laser sailing space-crafts.

In most proposals for interstellar travel, the space-crafts, it is argued, would travel at a few percentages of the speed limit $+c$ of light, as it is measured by one second, in free vacuum space and time. Therefore, it would seem that these cumulative kinematic and dynamic effects; and these cumulative thermo kinematic and dynamic effects happening to the mass and the structure of the space-craft *would not come into play*. Yet this conclusion is very much mistaken.

Since, there would have resulted mass structural damage to the entire space-craft, and these cumulative thermo dynamic effects, as experienced by this space-craft and upon any living observers within this space-craft. Both the space-craft, and their observers would be destroyed. And therefore, interstellar travel, even at a few percentages of the speed limit $+c$ of light, is physically impractical for any living observers, and also for their space-craft.

It is also our conclusion that these unique kinds of space-crafts, such as laser sailing space-crafts, have a propulsion system, a laser, would also experience mass structural damage to the entire space-craft—and whereas, to then make the space-craft physically useless for further interstellar travel.

Since, the mass structural damage to the space-craft would occur because to the cumulative thermo kinematic and dynamic effects. And which are caused by the instantaneous force: $+F_v$, *impacting*, instantaneously, constantly, and continuously: upon the mass structure, or upon the mass of the spacecraft—by its original *force thrust engine* upon the spacecraft—over an extended time period.[3]

Notes

1   In fact it is more than just a "paradox", it is in actuality an "anomaly" of the special relativity theory. And thus, this "anomaly" of the twin paradox, or of the identical clock paradox, could not be resolved by the special relativity theory. Because, these "paradoxes" were, in actuality, derived from the basic principles of the special relativity theory.

2   The special relativity theory predicts that it is theoretically feasible for a unique kind of space-craft, to approach the uniform vector velocity limit $+\underline{c}$, as it is measured by one second. Or to approach the uniform inertial scalar speed limit $+\underline{c}$ of light, as it is measured by one second.

And this is the case, provided, if there were enough fuel, or force to uniformly accelerate displace this unique kind of spacecraft, in free vacuum space and time. And some proponents of the special relativity theory have argued, or concluded, and confirmed, that there is not enough fuel, or force, within the entire cosmos to accomplish these events. Because, as the proponents of the special relativity theory have argued: that the mass of the spacecraft would grow to be "infinite mass". Which is silly.

Since, they have concluded, or inferred, that there would happen an instantaneous and constant "mass-increase" for this unique kind of spacecraft. Or for that matter, for any kind of mass, which would become "infinite mass". However, I reject both of these conclusions, or arguments, as derived from the special relativity theory. Since, it is true that the mass of the spacecraft would increase: it would have "imploded into itself". And yet, the mass of the spacecraft would have *drastically decreased* once the spacecraft begins a small, or a large approach to the uniform vector velocity limit +c, as it is measured by one second.

3   Or begin a small, or a large approach to the scalar speed limit $+\underline{c}$ of light, as it is measured by one second. in free vacuum space and time.

These events are caused by the instantaneous dynamic actions, as well as by the kinematic actions, of an applied force: $+F_v$—ie., of the "force-thrust" of the spacecraft's rocket engine.

The spacecraft, by then, would have "exploded outwards" from itself into being very hot fragments of mass and energy.

Although, in being more precise and objective, the spacecraft's propulsion engine, perhaps, would have been instantaneously "imploded/exploded". And thus, destroying the propulsion engine, and also, destroying the spacecraft. Or seriously damaging the spacecraft, and also killing any living observers within the spacecraft.

And yet, our new kinematic and dynamic scenarios, and our new electro-thermo kinematic and dynamic scenarios—is very different than what the special relativity theory would predict what would happen to this unique spacecraft. The proponents of the special relativity theory would argue that the spacecraft would approach "infinite mass", if the spacecraft approached, a "large approach" to the uniform scalar speed limit $+\underline{c}$, of light, as it is measured by one second, in free vacuum space and time. Which is silly. And the twin of this kind of spacecraft, in regards to its time, would be "zero-time", which is "dead-time", ie., the twin on this kind of spacecraft would have been long dead. And where the other twin safe on Earth would be alive, an "alive-time".

See bibliography. *Interstellar Migration and the Human Experiences*. And: *Starsailing: Solar Sails and the Inter-Stellar Travels*. And: *Time and the Space Traveller*.

APPENDIX D

ON TEMPORAL MOTIVE DISPLACEMENTS

-a-

There is a remarkable problem in the science of physics in regards to the uniform motive displacements for any cosmic masses, or for any cosmic phenomenon. Whether on the macro-scale of Nature, or on the micro scale of Nature: is their "uniform temporal-motive displacements", for these same natural phenomenon, or for these same cosmic masses.

Whereas, with respect to these natural phenomenon uniform motive displacements—or with respect to these cosmic masses uniform motive displacements from a position (a) to another position (b). By the dynamic actions of an applied force: $+F_v$, within a time interval t_b. Or of a cosmic mass having a uniform inertial velocity displacement, from the same positions (a) to (b), instantaneously, at the end of a time interval t_b. We then can describe and explain any natural phenomenon, or any cosmic mass, as possessing a "uniform temporal-motive displacements" *with respect to its uniform motive displacements*.

This fact of the "uniform temporal-motive displacements" for any natural phenomenon, of for any cosmic masses—in free vacuum space and time, should not be surprising. Since, many ancient people had observed and studied the uniform motion displacements of the Heavenly Bodies for thousands of years—and their observations, and their studies were made for practical reasons. Such as, for their constructions of time devices, and for their construction of calendar systems. And for other reasons, such as, for their religious and state affairs, and for devinational methods.

What these ancient peoples had observed, and studied, when they viewed or observed the Heavenly Bodies—and of the entire cosmos, as a whole, was a natural time device. Or a natural time machine, as based upon their understanding of the orderly and regular uniform motion displacements of these same Heavenly Bodies. And which these ancient people soon put into practical uses in developing their time devices, and their calendar systems. Such as, they knew that the regular uniform motions of the moon, of the sun, could be used by them in their development of new time devices, and calendar systems—for their daily affairs, and for their religious, and state affairs on the Earth. And it was only a short step for them to consider the entire Heavenly Bodies, and the entire cosmos, as a whole, to be some kind of natural time device, or a natural time machine.[1] As based upon their understanding of the uniform motion properties of these same Heavenly Bodies, and of the entire cosmos taken as a whole. And we can now conclude, with hindsight, that they were not far from the mark.

In the year 1582, Renaissance Italy, the young Galilie Galileo, had observed the uniform motion displacements properties of a swinging lamp, which was attached to a long cord. And Galileo "timed" the uniform motion displacement of the swinging lamp by using his heart beat. He had recognized that this phenomena of the "pendulum" could be used in the construction of a pendulum clock. And eventually, Christian Huyghens, in Holland, in the 17th century has so constructed the first successful working pendulum clock, as based upon Galileo's original insights.

And what these ancient scientists may have foreseen, was perhaps, (as Galileo, and Huyghens, may also have foreseen), was a new science of physics of the *uniform temporal-motive displacements*. As which was derived from their physics of the uniform motive displacements for any moving natural phenomenon, or for any moving cosmic mass.[2] But since they did not have a well worked out physics of the uniform motive displacements for any natural phenomenon, or for any cosmic mass, they could not then proceed further with a new science of the uniform temporal-motive displacements.

And only later, when Isaac Newton had advanced his *Mathematical Principles of Natural Philosophy*, in the year 1687. And only when Albert

Einstein had overturned some of the Newtonian kinematics and dynamics, by his *Special Theory of Relativity*, in the year 1905. It is now that we have reached a most precise stage in the development of physics to be able to consider fully this new conception of *temporal-motive-kinematics, and temporal-motive-dynamics*: for the uniform temporal-motive displacements for any temporal-motive cosmic mass +m, in free vacuum space and time.[3]

This new temporal-motive kinematics and dynamics we will advance in this Appendix D. Which is in regards to the uniform-motive displacements for all uniform temporal-motive cosmic masses, in free vacuum. And it is our conclusion that this new science of temporal-motive physics will be able to successfully deal with any problems of temporal-motive kinematics, and temporal-motive dynamics. And which this solution is in regards to the uniform temporal-motive displacements for any temporal-motive cosmic masses, in free space and time.

-b-

Given the new classical formulas: +mv, (+mc = 0), and: +mv, (+mc = 0). And multiplying these formulas by the limiting factor: +1/c, we then have the new temporal-motive formulas: +m(v / c), +m(c / c) = +m(1), and: +m(v / c), +m(c / c) = +m(1). Where the terms in parenthesis are to be operationally performed first. And where the unit vector term: +1 is called the uniform limit vector of one temporal second.[4] And where the unit scalar term: +1 is called the uniform limit scalar of one temporal-second. And where the fractional formula: +v / c is called the uniform limited vector fractional temporal-second formula—with respect to the uniform limit temporal-second of one vector +c / c second: +1. And where the fractional formula: +v / c is called the uniform limited scalar fractional temporal-second formula. Again with respect to the uniform limit temporal-second of one scalar +c / c second: +1.

Given the new classical formulas: +mv̇ , +(mċ = 0), and +mv, +(mċ = 0). And multiplying these formulas by the limiting factor: +1 / c, we then have the new temporal-motive formulas: +m(v̇ / c) , +(mċ / c) =0, and +m(v̇ / c), and: $+(\underline{m}\dot{c} \,/\, c) = 0$. And the terms in parenthesis are to be operationally performed first.

And where the fractional formula: $\dot{v}\,/\,\underline{c}$ is the first derivative of the limited vector: $\dot{v}$, divided by the speed limit $\underline{c}$ term of light. And this formula is called the uniform limited vector fractional second. Again with respect to a zero term "0" that the term: $+\dot{v}$ may be.[5] And where the fractional formula: $+\dot{v}\,/\,\underline{c}$ is also the first derivative of the limited scalar term: $+\underline{\dot{v}}$ divided by the speed limit $\underline{c}$ of light, in free vacuum.

And placing these new temporal-motive formulas into limit notations, we now have:

$$\mathrm{Lim}_{\substack{v\to c}}+\underline{m}(v\,/\,\underline{c}) = +\underline{m}(c\,/\,\underline{c}) = +\underline{m}\cdot 1 \tag{D.0}$$

$$\mathrm{Lim}_{\substack{\underline{v}\to\underline{c}}}+\underline{m}(\underline{v}\,/\,\underline{c}) = +\underline{m}(\underline{1}) \tag{D.1}$$

And:

$$\mathrm{Lim}+\underline{m}(\dot{v}\,/\,\underline{c}) = +\underline{m}(\dot{c}\,/\,\underline{c}) = 0$$
$$\dot{v}\to\dot{c} = 0 \tag{D.2}$$

$$\mathrm{Lim}+\underline{m}(\underline{\dot{v}}\,/\,\underline{c}) = +\underline{m}(\underline{\dot{c}}\,/\,\underline{c}) = 0$$
$$\underline{\dot{v}}\to\underline{\dot{c}} \tag{D.3}$$

Where the formula (D.0), on the left hand side, is the new limited vector temporal-velocity-momentum displacement formula of one vector fractional second. And where the formula (D.0), on the right hand side, is the new vector-temporal-velocity momentum formula of one vector second.

Where the formula (D.1), on the left hand side, is the new limited scalar temporal-speed-momentum formula of one scalar fractional second. And where the formula (D.1), on the right hand side, is the new limit scalar temporal-speed momentum formula of one scalar second.

Where the formula (D.2), on the left hand side, the derivative of the term: v, is the new limited vector temporal-acceleration force formula of one derivative fractional second. And where the formula (D.2), on the right hand side, is the first derivative of the term: $+\underline{c}$, is the new limit vector temporal-acceleration-force formula of one derivative zero one second.

Where the formula (D.3), on the left hand side, is the first derivative of the term: $+\underline{\dot{v}}$, is the new limited scalar temporal-acceleration-force formula of one derived fractional second. And where the formula (D.3), on the right hand side, is the first derivative of the term: $+\underline{c}$, is the new limit scalar temporal-acceleration-force formula of one derived zero one second.

By using the four limiting factors: $(1-(\underline{u}/\underline{c})^2)^{+\frac{1}{2}}$, $(1-(\underline{u}/\underline{c})^2)^{-\frac{1}{2}}$, and $(1-(\underline{c}/\underline{c})^2)^{+\frac{1}{2}}$, and $(1-(\underline{c}/\underline{c})^2)^{-\frac{1}{2}}$—upon the formulas of (D.0) to (D.3), we will have the following new temporal-motive formulas:

$$+\underline{m}(v/\underline{c})/(1-(\underline{u}/\underline{c})^2)^{+\frac{1}{2}} = \text{tem}_1 \tag{D.4}$$

And

$$+\underline{m}(v/\underline{c})(1-(\underline{u}/\underline{c})^2)^{+\frac{1}{2}} = \text{tem}_2 \tag{D.5}$$

Where the temporal kinematic formula (D.4) is called the limited momentum formula for external *vector temporal-velocity momentum* displacement (of decrease/of increase) for the temporal-motive mass $\underline{m}$.

And where the temporal kinematic formula (D.5) is called the limited momentum formula for external *vector temporal-mass-momentum* displacements (of increase/of decrease), for the temporal motive mass $\underline{m}$. We also have:

$$(+\underline{m})(c/\underline{c})(+1-(\underline{c}/\underline{c})^2)^{+\frac{1}{2}} = 0 \tag{D.6}$$

And:

$$(+\underline{m})(c/\underline{c})/(+1-(\underline{c}/\underline{c})^2)^{+\frac{1}{2}} = 0 \tag{D.7}$$

Where the temporal kinematic formula (D.6) is called the limit momentum formula for external *vector temporal-velocity momentum* displacement (of zero decrease/of zero increase) for temporal-motive mass +$\underline{m}$.

And where the temporal kinematic formula (D.7) is called the limit momentum formula for external vector *temporal-velocity momentum* displacement (of zero increase/of zero decrease) for the temporal-motive mass +$\underline{m}$.

And now given the extra scalar temporal momentum formulas: +$\underline{mv}$, and +$\underline{mc}$, we also have the new scalar temporal-momentum formulas:

$$+\underline{m}(\underline{v}/\underline{c})/(1-(\underline{u}/\underline{c})^2)^{+\frac{1}{2}} = \text{tem}_3 \tag{D.8}$$

And:

$$+\underline{m}(\underline{v}/\underline{c})(1-(\underline{u}/\underline{c})^2)^{+\frac{1}{2}} = \text{tem}_4 \tag{D.9}$$

Where the temporal kinematic formula (D.8) is called the limited momentum formula for external scalar *temporal-speed momentum* displacement (of decrease/of increase) for the temporal motive mass +$\underline{m}$.

And where the temporal kinematic formula (D.9) is called the limited momentum formula for external scalar *temporal-mass momentum* (of increase/of decrease) for the temporal motive mass +$\underline{m}$. We also have:

$$+\underline{m}(c/\underline{c})/(1-(\underline{c}/\underline{c})^2)^{+\frac{1}{2}} = 0 \tag{D.10}$$

And:

$$+\underline{m}(\underline{c}/\underline{c})(1-(\underline{c}/\underline{c})^2)^{\frac{1}{2}} = 0 \tag{D.11}$$

Where the temporal kinematic formula (D.10) is called the limit momentum formula for external scalar *temporal-speed-momentum* displacement (of zero decrease/of zero increase) for the temporal-motive mass $+\underline{m}$.

And where the temporal kinematic formula (D.11) is called the limit momentum formula for external scalar *temporal-speed* momentum displacement (of zero increase/of zero decrease) for a temporal-motive mass $+\underline{m}$.

And we now have:

$$+\underline{m}(\dot{v}/\underline{c})/(1-(\dot{\underline{u}}/\underline{c})^2)^{\frac{1}{2}} = \text{tem}_5 \qquad\qquad (D.12)$$

And:

$$+\underline{m}(\dot{v}/\underline{c})(1-(\dot{\underline{u}}/\underline{c})^2)^{\frac{1}{2}} = \text{tem}_{5.1} \qquad\qquad (D.13)$$

And where the temporal dynamic formula (D.12) is called the limited force formula for external *temporal-acceleration-force* displacement (of decrease/of increase) for the temporal-motive mass $+\underline{m}$.

And where the temporal dynamic formula (D.13) is called the limited force formula for external vector *temporal-mass-force* displacement (of increase/of decrease) for a temporal-motive mass $+\underline{m}$. And we have:

$$+\underline{m}(\dot{\underline{v}}/\underline{c})/(1-(\dot{\underline{u}}/\underline{c})^2)^{\frac{1}{2}} = \text{tem}_{5.2} \qquad\qquad (D.14)$$

and:

$$+\underline{m}(\dot{\underline{v}}/\underline{c})(1-(\dot{\underline{u}}/\underline{c})^2)^{\frac{1}{2}} = \text{tem}_{5.3} \qquad\qquad (D.15)$$

And where the temporal dynamic formula (D.14) is called the limited force formula for *external scalar temporal-speed force* displacement (of decrease/of increase) for a temporal motive mass $+\underline{m}$.

And where the temporal dynamic formula (D.15) is called the limited force formula for *external scalar temporal-mass-force* displacement (of increase/of decrease) for a temporal-motive mass $+\underline{m}$. We now have:

$$+\underline{m}(\dot{c}/\underline{c})/(1-(\dot{c}/\underline{c})^2)^{\frac{1}{2}} = \text{tem}_{5.4} = 0 \qquad\qquad (D.16)$$

and:

$$+\underline{m}(\dot{c}/\underline{c})(1-(\dot{c}/\underline{c})^2)^{\frac{1}{2}} = \text{tem}_{5.5} = 0 \qquad\qquad (D.17)$$

And where the temporal dynamic formula (D.16) is called the limit force formula for *external vector temporal-acceleration-force* displacement (of zero decrease/of zero increase) for a temporal-motive mass $+\underline{m}$.

And where the temporal dynamic formula (D.16) is called the limit force formula for external vector *temporal-acceleration-force* displacement[6] (of zero increase/of zero decrease) for a temporal-motive mass $+\underline{m}$.

We now have:

$$+\underline{m}(\dot{c}/\underline{c})/(1-(\dot{c}/\underline{c})^2)^{\frac{1}{2}} = \text{tem}_{5.6} = 0 \qquad\qquad (D.18)$$

and:

$$+\underline{m}(\dot{c}/\underline{c})(1-(\dot{c}/\underline{c})^2)^{\frac{1}{2}} = \text{tem}_{5.7} = 0 \qquad\qquad (D.19)$$

Where the temporal scalar formula (D.18) is called the limit force formula for external scalar *temporal-speed-force* displacement (of zero decrease/of zero increase) for a temporal-motive mass $+\underline{m}$.

And where the temporal scalar formula (D.19) is called the limit force formula for external scalar *temporal-mass-force* displacement (of zero increase/of zero decrease) for a temporal-motive mass $+\underline{m}$.

In this section—b—we have not made use of any volume $\underline{V}$, nor density $\underline{d}$ formulas. In the conclusion of this section—b-, we will list the volume $\underline{V}$, and density $\underline{d}$ temporal-motive formulas, without using any kind of limit notations. The reader of this appendix D, will complete the job of arranging the $+\underline{V}$ temporal-motive volume formulas, and of the $+\underline{d}$ as temporal-motive density formulas.

We then have:

$\underline{V}(\underline{v}\,/\,\underline{c})$, $\underline{V}(\dot{\underline{v}}\,/\,\underline{c})$, $\underline{V}(v\,/\,\underline{c})$, $\underline{V}(\dot{v}\,/\,\underline{c})$ and we have:

$\underline{V}(\underline{v}\,/\,\underline{c})\,/\,(1-(\underline{u}\,/\,\underline{c})^2)^{+\frac{1}{2}}$, $\underline{V}(\dot{\underline{v}}\,/\,\underline{c})\,/\,(1-(\dot{\underline{u}}\,/\,\underline{c})^2)^{+\frac{1}{2}}$, and:

$\underline{V}(v\,/\,\underline{c})\,/\,(1-(\underline{u}\,/\,\underline{c})^2)^{+\frac{1}{2}}$, and: $\underline{V}(\dot{v}\,/\,\underline{c})\,/\,(1-(\underline{u}\,/\,\underline{c})^2)^{+\frac{1}{2}}$ And these formulas are limited temporal-motive volume formulas.

We then have:

$\underline{d}(\underline{v}\,/\,\underline{c})$, $\underline{d}(\dot{\underline{v}}\,/\,\underline{c})$, $\underline{d}(\dot{v}\,/\,\underline{c})$, we then have:

$\underline{d}(\underline{v}\,/\,\underline{c})(1-(\underline{u}\,/\,\underline{c})^2)^{+\frac{1}{2}}$, $\underline{d}(\dot{\underline{v}}\,/\,\underline{c})(1-(\dot{\underline{u}}\,/\,\underline{c})^2)^{+\frac{1}{2}}$, $\underline{d}(v\,/\,\underline{c})(1-(\underline{u}\,/\,\underline{c})^2)^{+\frac{1}{2}}$

and: $\underline{d}(\dot{v}\,/\,\underline{c})(1-(\dot{\underline{u}}\,/\,\underline{c})^2)^{+\frac{1}{2}}$. And these formulas are the limited temporal-motive-density formulas.

We then have:

$\underline{V}(\underline{c}\,/\,\underline{c})$, $\underline{V}(\dot{\underline{c}}\,/\,\underline{c})$, and we have:

$\underline{V}(\underline{c}\,/\,\underline{c})\,/\,(1-(\underline{c}\,/\,\underline{c})^2)^{+\frac{1}{2}}$, $\underline{V}(\dot{\underline{c}}\,/\,\underline{c})\,/\,(1-(\dot{\underline{c}}\,/\,\underline{c})^2)^{+\frac{1}{2}}$, $\underline{V}(c\,/\,\underline{c})^2)^{+\frac{1}{2}}$,

and: $\underline{V}(\dot{c}\,/\,\underline{c})\,/\,(1-(\dot{\underline{c}}\,/\,\underline{c})^2)^{+\frac{1}{2}}$. And these formulas are the limit temporal-motive-volume formulas.

We then have:

$\underline{d}(\underline{c}\,/\,\underline{c})$, $\underline{d}(\dot{\underline{c}}\,/\,\underline{c})$, $\underline{d}(c\,/\,\underline{c})$, we then have:

$\underline{d}(\underline{c}\,/\,\underline{c})(1-(\underline{c}\,/\,\underline{c})^2)^{+\frac{1}{2}}$, $\underline{d}(\dot{\underline{c}}\,/\,\underline{c})(1-(\dot{\underline{c}}\,/\,\underline{c})^2)^{+\frac{1}{2}}$, $\underline{d}(c\,/\,\underline{c})(1-(\underline{c}\,/\,\underline{c})^2)^{+\frac{1}{2}}$

and: $\underline{d}(\dot{c}\,/\,\underline{c})(1-(\dot{\underline{c}}\,/\,\underline{c})^2)^{+\frac{1}{2}}$. And these formulas are the limit temporal-motive-density formulas.

Where the term $\underline{V}$ is volume, and where the term: $\underline{d}$ is density, and where the term $\underline{c}$ is the speed of light in vacuum. and the term: c is the vector speed of light in vacuum,

$$-c-$$

For a change of mind, I will include the temporal-motive displacement formulas for density, and for volume. We have:

$$\underline{m} / \underline{d}(\underline{v} / \underline{c}) = \underline{V}(\underline{v} / \underline{c}) \qquad\qquad (D.20)$$

$$\underline{m} / \underline{d}(\underline{c} / \underline{c}) = \underline{V}(\underline{1}) \qquad\qquad (D.21)$$

And:

$$\mathrm{Lim} V(\underline{v} / \underline{c}) = \underline{V}(\underline{c} / \underline{c}) = \underline{V}(\underline{1}) \qquad\qquad (D.22)$$
$$\underline{v} \to \underline{c}$$

And:

$$\underline{m} / \underline{d}(v / \underline{c}) = V(v / \underline{c}) \qquad\qquad (D.23)$$

$$\underline{m} / \underline{d}(c / \underline{c}) = v \cdot (1) \qquad\qquad (D.24)$$

$$\mathrm{Lim}\underline{V}(v / \underline{c}) = V(c / \underline{c}) = \underline{V} \cdot (1) \qquad\qquad (D.25)$$
$$v \to c$$

Where $\underline{V}$ is volume, and where $\underline{d}$ is density, and while $+\underline{m}$ is mass. We now have:

$$\underline{m} / \underline{V}(\underline{v} / \underline{c}) = \underline{d}(\underline{v} / \underline{c}) \qquad\qquad (D.26)$$

$$\underline{m} / \underline{V}(\underline{c} / \underline{c}) = \underline{d}(\underline{c} / \underline{c}) = \underline{d}(\underline{1}) \qquad\qquad (D.27)$$

And:

$$\text{Lim}\underline{d}(\underline{v} / \underline{c}) = \underline{d}(\underline{c} / \underline{c})\text{---}\underline{d}(\underline{1}) \qquad \text{(D.28)}$$
$$\underline{v} \rightarrow \underline{c}$$

And:

$$\underline{m} / \underline{V}(v / \underline{c}) = \underline{d}(v / \underline{c}) \qquad \text{(D.29)}$$

$$\underline{m} / \underline{V}(c / \underline{c}) = \underline{d}(1) \qquad \text{(D.30)}$$

And:

$$\text{Lim}\underline{d}(v / \underline{c}) = \underline{d}(c / \underline{c}) = \underline{d}(1) \qquad \text{(D.31)}$$
$$+v \rightarrow +c$$

The terms $\underline{v}$, and $\underline{c}$, and $\underline{1}$, are scalar terms, representing limited scalar speed $\underline{v}$, and representing scalar limit speed $\underline{c}$ of light in free vacuum. And the term 1 represents one limit scalar second. And the terms $\underline{d}$, and $\underline{V}$, and $\underline{m}$, are also scalar terms, representing one limited density $\underline{d}$, and while the scalar term $\underline{V}$ represents one limited volume. And while the scalar term $\underline{m}$ represents one limited mass.

Where the terms v, c, and 1, represents one limited vector velocity v displacement term. And the term c represents one limit vector velocity c displacement term of vector light, in free vacuum. And where the term 1 represents one limit vector second.

We have:

$$\underline{m} / \underline{d}(\dot{\underline{v}} / \underline{c}) = \underline{V}(\dot{\underline{v}} / \underline{c}) \qquad \text{(D.32)}$$

$$(+\underline{m} / \underline{d})(\dot{\underline{c}} / \underline{c}) = V(0 / \underline{c}) = 0 \qquad \text{(D.33)}$$

then:

$$\text{Lim}\underline{V}(\dot{\underline{v}} / \underline{c}) = \underline{V}(\dot{\underline{c}} / \underline{c}) = \underline{V}(0 / \underline{c}) = 0$$
$$\dot{\underline{v}} \rightarrow (\dot{\underline{c}} = 0) \qquad \text{(D.34)}$$

And we have:

$$\underline{m} / \underline{d}(\dot{v} / \underline{c}) = \underline{V}(\dot{v} / \underline{c}) \tag{D.35}$$

$$\underline{m} / \underline{d}(\dot{c} / \underline{c}) = \underline{V}(\dot{c} / \underline{c}) = \underline{V}(0 / \underline{c}) = 0 \tag{D.36}$$

And:

$$\begin{aligned} &\operatorname{Lim}\underline{V}(\dot{v} / \underline{c}) = \underline{V}(\dot{c} / \underline{c}) = \underline{V}(\dot{1}) = 0 \\ &+\underline{\dot{v}} \rightarrow +\underline{\dot{c}} \end{aligned} \tag{D.37}$$

Regarding the terms $\dot{v}$, and $\dot{1}$, are the first derivative[7] of the limited acceleration displacement term: $+\dot{v}$. And the limit term: $+\underline{\dot{c}}$, is the first derivative of the limit vector velocity displacement term: $+c$. And the first derivative of the limit temporal-motive displacement term: $+1$, is $(+\dot{1} = 0)$.

And we now have:

$$\underline{m} / \underline{V}(\dot{v} / \underline{c}) = \underline{d}(\dot{v} / \underline{c}) \tag{D.38}$$

$$\underline{m} / \underline{V}(\dot{c} / \underline{c}) = \underline{d}(\dot{c} / \underline{c}) = \underline{d}(\dot{1}) = 0 \tag{D.39}$$

then:

$$\begin{aligned} &\operatorname{Lim}\underline{d}(\dot{v} / \underline{c}) = \underline{d}(\dot{c} / \underline{c}) = \underline{d}(\dot{1}) = 0 \\ &\underline{\dot{v}} \rightarrow \underline{\dot{c}} = 0 \end{aligned} \tag{D.40}$$

We will now give the interpretations of all formulas, from formula (D.20) to formula (D.40). The division of the density $\underline{d}$ into the mass $\underline{m}$, of formula D.20), is the new formula of limited temporal-motive volume-momentum displacement formula.[8] And the formula (D.21) is the division of the density mass $\underline{m}$, of the same formula, is the new formula of limit temporal-motive volume-momentum displacement formula. The limit (D.22) is the limit of the limited formulas, on the left hand side, of this limit notation, approaches, but never equals the limit formula, on the right hand side of this limit (D.22). This limit (D.22), in regards to their temporal-motive displacements, signifies, that the limited expression, or formula $\underline{V}(\underline{v} / \underline{c})$—which can be defined as limited temporal-motive volume-momentum displacement. But the limited temporal-motive volume-momentum formula: $\underline{V}(\underline{v} / \underline{c})$, which never equals the limit temporal-motive volume-momentum displacement formula: $\underline{V}(\underline{v} / \underline{c}) = \underline{V}(\underline{1})$. Where the term: $\underline{1}$ is of one scalar second. And while the term $(\underline{v} / \underline{c}$ is of a fractional second, a scalar term, which also varies, with respect to the varying volume of this limit notation (D.22). The term: $\underline{v}$, a scalar term, also varies, but never achieves the limit term: $\underline{c}$, which is the term, or constant for the speed of light, in free vacuum.

The division of the density $\underline{d}$ into the mass $\underline{m}$, of formula (D.23), is the new formula of limited vector temporal-motive volume-momentum displacement formula. Which the term: $(v / \underline{c})$ is of fractional vector second, while the vector term: v varies but never equals the limit vector term c, with respect to volume. The term $(c / \underline{c})$ is of one vector second, 1. And we can say, that the term: $\underline{V}(c / \underline{c})$ is the limit vector temporal-motive volume-momentum displacement formula of one vector second.

And we have the limit notation (D.25), we have:

$$\text{Lim}\underline{V}(v / \underline{c}) = \underline{V}(c / \underline{c}) = \underline{V}\cdots (1) \qquad\qquad (D.25)$$
$$v \rightarrow c$$

The new vector temporal-motive volume-momentum displacement formula, which is limited, never equals the new limit c, which is the limit temporal-motive volume-momentum displacement formula—as the limit notation signifies.

The terms: $v / \underline{c}$ is of one vector variable fractional second. Both terms are in respect to the varying volume, with respect to a mass $\underline{m}$, and its density $\underline{d}$.

Where $\underline{V}$ is volume, which varies, and where $\underline{d}$ is density, which also varies, and while $\underline{m}$ is mass, which also varies, according to applied forces upon the mass $\underline{m}$, or perhaps due to the natural forces upon the mass $\underline{m}$.

And the term $\underline{v}$ is speed, a scalar term, and while +v is vector velocity, and while $\underline{c}$ is a scalar speed limit of light in free vacuum. And while the term c is vector velocity, a limit, and perhaps a constant in its own right, in regards to all of the other constants of nature. The term c is an idealized term, but it still have its function in nature, or in our explanation of nature.

We now have:

$$\underline{m} / \underline{V}(\underline{v} / \underline{c}) = \underline{d}(\underline{v} / \underline{c}) \tag{D.26}$$

The division of the volume term $\underline{V}$ into the mass term $\underline{m}$, is the new formula for limited temporal-motive density-momentum displaced formula. Where the term: $\underline{v} / \underline{c}$ is of one fractional second with respect to the term $\underline{c} / \underline{c} = \underline{1}$, which is limit temporal-motive density-momentum displaced formula, which is one scalar second.

The division of the volume $\underline{V}$ into mass $\underline{m}$, as the formula (D.26) indicates, results in density. (D.26) is explained as the division of volume into mass is density, which the terms $\underline{v} / \underline{c}$ are of limited yet fractional temporal-motive density-momentum displaced formula, in regards to its density, and its mass.

The division of the volume $\underline{V}$ into mass $\underline{m}$, as the formula (D.27) indicates, results in density. (D.27) is explained as the division of volume into mass is density, which the terms $\underline{c} / \underline{c}$ are of a limit, yet constant, temporal-motive density-momentum displaced formula, in regards to its density, and its mass.[9] We have:

$$\text{Lim}\underline{d}(\underline{v} / \underline{c}) = \underline{d}(\underline{c} / \underline{c}) = \underline{d}(\underline{1}) \tag{D.28}$$
$$\underline{v} \rightarrow \underline{c}$$

The term $\underline{c} / \underline{c} = \underline{1}$ is of one scalar limit temporal-motive density-momentum displaced formula. And where the term $\underline{v} / \underline{c}$ is one scalar limited temporal-motive density-momentum displaced formula. Where the term $\underline{1}$ is of one scalar second. and where the term $\underline{v} / \underline{c}$ is of one limited fractional second. The term $\underline{1}$ is a limit, or constant in its own right.

We now have the limited formula (D.29), as follows.

$$\underline{m} / \underline{V}(v / \underline{c}) = \underline{d}(v / \underline{c}) \tag{D.29}$$

The division of the volume $\underline{V}$ into the mass $\underline{m}$, of formula (D.29), is the new formula of limited vector temporal-motive density-momentum displaced formula. And which the term: $v / \underline{c}$ is of a fractional vector second, while the vector term: v varies but never equals the limit vector term: c, with respect to density.

We now have:

$$\underline{m} / \underline{V}(c / \underline{c}) = \underline{d} \cdot (1) \tag{D.30}$$

The division of the volume $\underline{V}$ into the mass $\underline{m}$, of formula (D.30), is the new formula of limit vector velocity temporal-motive density-momentum displaced formula. And which the term: $c / \underline{c}$ is one vector second, and while the formula on the right hand side of (D.30) is also a limit vector velocity temporal-motive density-momentum displaced formula. The terms $c / \underline{c}$ and 1 are one vector second.

And we have the limit notation (D.31), we have:

$$\underset{v \to c}{\text{Lim}} \underline{d}(v / \underline{c}) = \underline{d}(c / \underline{c}) = \underline{d} \cdots (1) \tag{D.31}$$

The new vector temporal-motive density-momentum displacement formula, which is limited, never equals the new limit c, which is the limit temporal-motive density-momentum displacement formula—as the limit notation signifies.

And we have:

$$\underline{m} / \underline{d}(\dot{v} / \underline{c}) = \underline{V}(\dot{v} / \underline{c}) \tag{D.32}$$

The division of the density term $\underline{d}$ into the mass term $\underline{m}$ is the new formula for limited first derivative temporal-motive volume-force displaced formula. Where the term: $\dot{v}/c$ is of the first derivative of one fractional second—with respect to the term: $c/c = 1$, which is the limit of temporal-motive volume displaced formula, of one scalar second.

We now have:

$$\underline{m}/\underline{d}(\dot{c}/\underline{c}) = \underline{V}(0/\underline{c}) = 0 \tag{D.33}$$

The division of the density term $\underline{d}$ into the mass term $\underline{m}$ is the new formula for limit first derivative of the temporal-motive volume-force displaced formula. Where the term: $\dot{c}/\underline{c} = 0$, is of the first derivative of one zero second, with respect to the term: $c/\underline{c} = 1$, which is the limit temporal-motive volume displaced formula, of one scalar second.

We have:

$$\operatorname{Lim}\underline{V}(\dot{v}/\underline{c}) = \underline{V}(\dot{c}/\underline{c}) = \underline{V}(0/\underline{c}) = 0$$
$$\dot{v} \to (\dot{c} = 0) \tag{D.34}$$

The term $\dot{v}/\underline{c}$ is the first derivative of one scalar limited temporal-motive volume-force displaced formula. And where the term $\dot{c}/\underline{c} = 0$, is the first derivative of one scalar limit temporal-motive volume-force displaced formula. And the total limit is zero (0).

And we have:

$$\underline{m}/\underline{d}(\dot{v}/\underline{c}) = \underline{V}(\dot{v}/\underline{c}) \tag{D.35}$$

The division of the density term $\underline{d}$ into the mass term $\underline{m}$ is the new formula for vector limited first derivative temporal motive volume-force displaced formula. Where the term: $\dot{v}/c$ is of the vector first derivative of one vector fractional second. With respect to the term $c/\underline{c} = 1$, which is the of the temporal-motive volume displaced formula, of one scalar second.

We now have:

$$\underline{m}/\underline{d}(\dot{c}/\underline{c}) = \underline{V}(\dot{c}/\underline{c}) + V(0/\underline{c}) = 0 \tag{D.36}$$

The division of the density term $\underline{d}$ into the mass term $\underline{m}$ is the new formula for vector limit first derivative temporal-motive volume-force displaced formula. Where the term: $\underline{\dot{c}} / \underline{c} = 0$, is of the vector first derivative of one zero vector second.

We have:

$$\text{Lim}(+\underline{m}/\underline{d}) / (\dot{v}/\underline{c}) = \underline{V}(\dot{c}/\underline{c}) = \underline{V}(0/\underline{c}) = 0$$
$$\dot{v} \rightarrow (\dot{c} = 0) \tag{D.37}$$

The term $\dot{v}/\underline{c}$ is the first derivative of one vector limited temporal-motive volume-force displaced formula. And where the term $\dot{c}/\underline{c} = 0$, is the first derivative of one vector limit temporal-motive volume-force displaced formula. And the total limit is zero (0).

And we have:

$$\underline{m}/\underline{V}(\dot{v}/\underline{c}) = \underline{d}(\dot{v}/\underline{c}) \tag{D.38}$$

The division of the volume term $\underline{V}$ into the mass term $\underline{m}$ is the new formula for vector limited first derivative temporal motive density-force displaced formula. Where the term: $\dot{v}/\underline{c}$ is of the vector first derivative of one vector fractional second. With respect to the term: $\underline{c}/\underline{c} = \underline{1}$, which is the temporal-motive density displaced formula, or one scalar second.

We now have:

$$\underline{m}/\underline{V}(\dot{c}/\underline{c}) = \underline{d}(\dot{c}/\underline{c}) = \underline{d}(0/\underline{c}) = 0 \tag{D.39}$$

The division of the volume term $\underline{V}$ into the mass term $\underline{m}$ is the new formula for vector limit derivative of temporal-motive density-force displaced formula. Where the term: $\dot{c}/\underline{c} = 0$, is of the vector first derivative of one zero second.

We have:

$$\text{Lim}+\underline{m}/\underline{V})(\underline{\dot{v}}/\underline{c}) = \underline{d}(\underline{\dot{c}}/\underline{c}) = \underline{d}(0/\underline{c}) = 0$$
$$\dot{v} \rightarrow \dot{c} = 0) \tag{D.40}$$

The term $\dot{v}/\underline{c}$ is the first derivative of one vector limited temporal-motive density-force displaced formula. And where the term: $\dot{c}/\underline{c}=0$, is the first derivative of one vector limit temporal-motive density-force displaced formula. And the total limit is zero (0).

-d-

In regards to the volume term: $\underline{V}$, and the density term $\underline{d}$, and the limiting factors: $(1-(\underline{u}/\underline{c})^2)^{+\frac{1}{2}}$ and $(1-(\underline{u}/\underline{c})^2)^{-\frac{1}{2}}$, and $(1-(\underline{\dot{u}}/\underline{c})^2)^{+\frac{1}{2}}$ and $(1-(\underline{\dot{u}}/\underline{c})^2)^{-\frac{1}{2}}$, and $(1-(\underline{c}/\underline{c})^2)^{+\frac{1}{2}}$, and $(1-(\underline{c}/\underline{c})^2)^{-\frac{1}{2}}$, and $(1-(\underline{\dot{c}}/\underline{c})^2)^{+\frac{1}{2}}$, and $(1-(\underline{\dot{c}}/\underline{c})^2)^{-\frac{1}{2}}$. We will then give the complete kinematic and dynamic formulas.

Given the formulas (D.20) and (D.26), we have:

$$\underline{V}(\underline{v}/\underline{c})/(1-(\underline{u}/\underline{c})^2)^{+\frac{1}{2}} = \text{tem}_6 \qquad\qquad (D.41)$$

Where the formula (D.41) is the new limited kinematic scalar temporal speed volume momentum formula (of decrease/of increase) displacement, of one fractional second.

And we have formula (D.42):

$$\underline{d}(\underline{v}/\underline{c})(1-(\underline{u}/\underline{c})^2)^{+\frac{1}{2}} = \text{tem}_7 \qquad\qquad (D.42)$$

Where the formula (D.42) is the new limited temporal kinematic scalar temporal density momentum formula (of increase/of decrease) displacement, of one fractional second.

And given the formulas (D.21) and (D.27), we have:

$$\underline{v}(1)/(+1-(\underline{c}/\underline{c})^2)^{+\frac{1}{2}} = \text{tem}_{7.1} = 0 \qquad\qquad (D.43)$$

Where the formula (D.43) is the new limit kinematic scalar zero temporal volume speed momentum formula (of zero decrease/of zero increase) displacement, of one zero fractional second.

And we have:

$$\underline{d}(\underline{1})(1-(\underline{c}/\underline{c})^2)^{+\frac{1}{2}} = \text{tem}_{7.2} = 0 \tag{D.44}$$

Where the formula (D.44) is the new limit kinematic scalar zero temporal density-momentum formula (of zero increase/of zero decrease) displacement, of one zero second.

Given the formulas: $\underline{V}(\underline{v}/\underline{c})/(1-(\underline{u}/\underline{c})^2)^{+\frac{1}{2}}$, and: $\underline{V}(\underline{c}/\underline{c})/(1-(\underline{c}/\underline{c})^2)^{+\frac{1}{2}}$, we have the limit notation:

$$\begin{aligned}\text{Lim}\,\underline{V}(\underline{v}/\underline{c})/(1-(\underline{u}/\underline{c})^2)^{+\frac{1}{2}} &= \underline{V}(\underline{c}/\underline{c})/(1-(\underline{c}/\underline{c})^2)^{+\frac{1}{2}} = 0\\ \underline{v}&\to\underline{c}\\ \underline{u}&\to\underline{c}\end{aligned} \tag{D.45}$$

Given the formulas (D.23) and (D.29), we have:

$$\underline{V}(\underline{v}/\underline{c})/(1-(\underline{u}/\underline{c})^2)^{+\frac{1}{2}} = \text{tem}_8 \tag{D.46}$$

Where the formula (D.46) is the new limited dynamic vector velocity volume-momentum formula (of decrease/of increase) displacement, of one fractional second.

And we have the formula (D.47):

$$\underline{d}(\underline{v}/\underline{c})(1-(\underline{u}/\underline{c})^2)^{+\frac{1}{2}} = \text{tem}_9 \tag{D.47}$$

Where the formula (D.47) is the new limited dynamic vector velocity density-momentum formula (of increase/of decrease) displacement, of one fractional second.

And we have formula: (D.48):

$$\underline{V}(c/\underline{c})/(1-(\underline{c}/\underline{c})^2)^{+\frac{1}{2}} = \text{tem}_{10} \tag{D.48}$$

Where the formula (D.48) is the new limit dynamic vector velocity volume-momentum formula (of zero decrease/of zero increase), of one zero fractional second.

And we have formula (D.49):

$$\underline{d}(c/\underline{c})(1-(\underline{c}/\underline{c})^2)^{+\frac{1}{2}} = \text{tem}_{11} = 0 \tag{D.49}$$

Where the formula (D.49) is the new limit dynamic vector velocity density-momentum formula (of zero increase/of zero decrease) displacement, of one zero fractional second.

Given the formulas: (D.46), (D.47), (D.48), and (D.49), we then have the two limit notations:

$$\underset{\substack{v \to c \\ \underline{u} \to \underline{c}}}{\text{Lim}}\underline{V}(v/\underline{c})/(1-(\underline{u}/\underline{c})^2)^{+\frac{1}{2}} = \underline{V}(c/\underline{c})/(1-(\underline{c}/\underline{c})^2)^{+\frac{1}{2}} = 0 \tag{D.50}$$

And we have:

$$\underset{\substack{v \to c \\ \underline{u} \to \underline{c}}}{\text{Lim}}\underline{d}(v/\underline{c})(1-(\underline{u}/\underline{c})^2)^{+\frac{1}{2}} = \underline{d}(c/\underline{c})(1-(\underline{c}/\underline{c})^2)^{+\frac{1}{2}} = 0 \tag{D.51}$$

Given the formulas: $\underline{V}(\dot{v}/\underline{c})$, $\underline{V}(\dot{v}/\underline{c})$, $\underline{d}(\dot{v}/\underline{c})$, $d(\dot{v}/\underline{c})$, and: $\underline{V}(\dot{c}/\underline{c})$, $\underline{V}(\dot{c}/\underline{c})$, $\underline{d}(\dot{c}/\underline{c})$, and $\underline{d}(\dot{c}/\underline{c})$. And the limiting factors: $(1-(\underline{\dot{u}}/\underline{c})^2)^{+\frac{1}{2}}$, and

$(1-(\underline{\dot{u}}/\underline{c})^2)^{-\frac{1}{2}}$, and: $(1-(\underline{\dot{c}}/\underline{c})^2)^{+\frac{1}{2}}$, and $(1-(\underline{\dot{c}}/\underline{c})^2)^{-\frac{1}{2}}$. We now have the following formulas:

$$\underline{V}(\underline{\dot{v}}/\underline{c})/(1-(\underline{\dot{u}}/\underline{c})^2)^{+\frac{1}{2}} = \text{tem}_{12} \tag{D.52}$$

Where the formula (D.52) is the new limited dynamic scalar temporal-volume acceleration-force formula (of decrease/of increase) displacement.

And we have:

$$\underline{d}(\underline{\dot{v}}/\underline{c})(1-(\underline{\dot{u}}/\underline{c})^2)^{+\frac{1}{2}} = \text{tem}_{13} \tag{D.53}$$

Where the formula (D.53) is the new limited dynamic scalar temporal-density-force formula (of increase/of decrease) displacement.

And we have the following formulas:

$$\underline{V}(\underline{\dot{c}}/\underline{c})/(1-(\underline{\dot{c}}/\underline{c})^2)^{+\frac{1}{2}} = \text{tem}_{14} = 0 \tag{D.54}$$

Where the formula (D.54) is the new limit dynamic scalar temporal-volume-acceleration-force formula (of decrease/of increase) displacement.

And we have:

$$\underline{d}(\underline{\dot{c}}/\underline{c})(1-(\underline{\dot{c}}/\underline{c})^2)^{+\frac{1}{2}} = \text{tem}_{15} = 0 \tag{D.55}$$

Where the formula (D.55) is the new limit dynamic scalar temporal-density-force formula (of increase/ of decrease) displacement.

And placing the formulas (D.52), (D.53), and (D.54), and (D.55) into limit notations we have:

$$\text{Lim}\,\underline{V}(\underline{\dot{v}}/\underline{c})/(1-(\underline{\dot{u}}/\underline{c})^2)^{+\frac{1}{2}} = \text{tem}_{16} = 0$$
$$\underline{\dot{u}} \to \underline{\dot{c}} \tag{D.56}$$
$$\underline{\dot{v}} \to \underline{\dot{c}} = 0$$

And:

$$\text{Lim}\underline{d}(\underline{\dot{v}}/\underline{c})(1-(\underline{\dot{u}}/\underline{c})^2)^{+\frac{1}{2}} = \underline{d}(\underline{\dot{c}}/\underline{c})(1-(\underline{\dot{c}}/\underline{c})^2)^{+\frac{1}{2}} = 0$$
$$\underline{\dot{u}} \to \underline{\dot{c}} = 0 \qquad\qquad\qquad\qquad\qquad\qquad (D.57)$$
$$\underline{\dot{v}} \to \underline{\dot{c}}$$

We now have the following formulas:

$$\underline{V}(\underline{\dot{v}}/\underline{c})/(1-(\underline{\dot{u}}/\underline{c})^2)^{+\frac{1}{2}} = \text{tem}_{17} \qquad\qquad\qquad (D.58)$$

Where the formula (D.58) is the new limited dynamic vector temporal-volume acceleration-force formula (of decrease/of increase) displacement.

We now have the following formula:

$$\underline{d}(\underline{\dot{v}}/\underline{c})(1-(\underline{\dot{u}}/\underline{c})^2)^{+\frac{1}{2}} = \text{tem}_{18} \qquad\qquad\qquad (D.59)$$

Where the formula (D.59) is the new limited dynamic vector temporal-density-force formula (of increase/ of decrease) displacement.

And we have:

$$\underline{V}(\underline{\dot{c}}/\underline{c})/(1-(\underline{\dot{c}}/\underline{c})^2)^{+\frac{1}{2}} = \text{tem}_{19} = 0 \qquad\qquad\qquad (D.60)$$

Where the formula (D.60) is the new limit dynamic vector temporal-volume-acceleration-force formula (of decrease/of increase) displacement, of one zero second.

And we have:

$$\underline{d}(\underline{\dot{c}}/\underline{c})(1-(\underline{\dot{c}}/\underline{c})^2)^{+\frac{1}{2}} = \text{tem}_{20} = 0 \qquad\qquad\qquad (D.61)$$

Where the formula (D.61) is the new limit dynamic vector temporal-density-force formula (of increase/of decrease) displacement, of one zero second.

And placing formulas (D.58), (D.59), and D.60), and (D.61) into limit notations, we have:

$$\begin{aligned}&\operatorname{Lim}\underline{V}(\dot{v}/\underline{c})/(1-(\underline{\dot{u}}/\underline{c})^2)^{+\frac{1}{2}} = \underline{V}(\dot{c}/\underline{c})/(1-(\underline{\dot{c}}/\underline{c})^2)^{+\frac{1}{2}} = 0\\ &\dot{v} \to \underline{\dot{c}} = 0\\ &\underline{\dot{u}} \to \underline{\dot{c}} = 0\end{aligned} \qquad (D.62)$$

And we have:

$$\begin{aligned}&\operatorname{Lim}\underline{d}(\dot{v}/\underline{c})(1-(\underline{\dot{u}}/\underline{c})^2)^{+\frac{1}{2}} = \underline{d}(\dot{c}/\underline{c})(1-(\underline{\dot{c}}/\underline{c})^2)^{+\frac{1}{2}} = 0\\ &\dot{v} \to \dot{c} = 0\\ &\underline{\dot{u}} \to \underline{\dot{c}} = 0\end{aligned} \qquad (D.63)$$

-e-

In this section—e—we will deal with the temporal-motive kinetic energy formulas:[8]

$$\frac{1}{2}\underline{m}(\underline{v}/\underline{c})^2,\ \frac{1}{2}\underline{m}(v/\underline{c})^2,\ \frac{1}{2}\underline{m}(\underline{c}/\underline{c})^2,\ \frac{1}{2}\underline{m}(c/\underline{c})^2 \qquad (D.64)$$

which are the limited and limit temporal-motive-mass kinetic energy formulas.

And we have:

$$\frac{1}{2}(\underline{V})(\underline{v}/\underline{c})^2,\ \frac{1}{2}(\underline{V})(v/\underline{c})^2,\ \frac{1}{2}(\underline{V})(\underline{c}/\underline{c})^2,\ \frac{1}{2}(\underline{V})(c/\underline{c})^2 \qquad (D.65)$$

which are the limited and limit temporal-motive-volume kinetic energy formulas.

And we have:

$$\frac{1}{2}\underline{d}(\underline{v}/\underline{c})^2,\ \frac{1}{2}\underline{d}(v/\underline{c})^2,\ \frac{1}{2}\underline{d}(\underline{c}/\underline{c})^2,\ \frac{1}{2}\underline{d}(c/\underline{c})^2 \tag{D.66}$$

which are the limited and limit temporal-motive-density kinetic energy formulas.

And we have:

$$\frac{1}{2}\underline{m}(\underline{\dot{v}}/\underline{c})^2,\ \frac{1}{2}\underline{m}(\dot{v}/\underline{c})^2,\ \frac{1}{2}\underline{m}(\underline{\dot{c}}/\underline{c})^2,\ \frac{1}{2}m(\dot{c}/\underline{c})^2 \tag{D.67}$$

which are the first derivative of the limited and limit temporal-motive-mass-kinetic energy formulas.

And we have:

$$\frac{1}{2}V(\underline{\dot{v}}/\underline{c})^2,\ \frac{1}{2}\underline{V}(\dot{v}/\underline{c})^2,\ \frac{1}{2}\underline{V}(\underline{\dot{c}}/\underline{c})^2,\ \frac{1}{2}\underline{V}(\dot{c}/\underline{c})^2 \tag{D.68}$$

which are the first derivative of the limited and limit temporal-motive-volume-kinetic energy formulas.

And we have:

$$\frac{1}{2}\underline{d}(\underline{\dot{v}}/\underline{c})^2,\ \frac{1}{2}\underline{d}(\dot{v}/\underline{c})^2,\ \frac{1}{2}\underline{d}(\underline{\dot{c}}/\underline{c})^2,\ \frac{1}{2}\underline{d}(\dot{c}/\underline{c})^2 \tag{D.69}$$

which are the first derivative of the limited and limit temporal-motive-density-kinetic energy formulas.

Placing these formulas from (D.64) to (D.69), we have:

$$\text{Lim} \frac{1}{2}\underline{m}(\underline{v}/\underline{c})^2 = \frac{1}{2}\underline{m}(\underline{c}/\underline{c})^2 = \frac{1}{2}\underline{m}(1)$$
$$\underline{v} \to \underline{c}$$

(D.70)

And:

$$\text{Lim} \frac{1}{2}\underline{m}(v/\underline{c})^2 = \frac{1}{2}\underline{m}(c/\underline{c})^2 = \frac{1}{2}\underline{m}(1)$$
$$v \to c$$

(D.71)

And:

$$\text{Lim} \frac{1}{2}\underline{V}(\underline{v}/\underline{c})^2 = \frac{1}{2}\underline{V}(\underline{c}/\underline{c})^2 = \frac{1}{2}\underline{V}(1)$$
$$\underline{v} \to \underline{c}$$

(D.72)

And:

$$\text{Lim} \frac{1}{2}\underline{V}(v/\underline{c})^2 = \frac{1}{2}\underline{V}(c/\underline{c})^2 = \frac{1}{2}\underline{V}(1)$$
$$v \to c$$

(D.73)

And:

$$\text{Lim} \frac{1}{2}\underline{d}(\underline{v}/\underline{c})^2 = \frac{1}{2}\underline{d}(\underline{c}/\underline{c})^2 = \frac{1}{2}\underline{d}(1)$$
$$\underline{v} \to \underline{c}$$

(D.74)

And:

$$\text{Lim} \frac{1}{2}\underline{d}(v/\underline{c})^2 = \frac{1}{2}\underline{d}(c/\underline{c})^2 = \frac{1}{2}\underline{d}(1)$$
$$v \to c$$

(D.75)

Where the term $\underline{m}$ is mass, the term $\underline{V}$ is volume, and the term $\underline{d}$ is density.

We will take the first derivative of the formulas (D.64) to (D.66).

And we have:

$$\text{Lim} \frac{1}{2}\underline{m}(\dot{\underline{v}}/\underline{c})^2 = \frac{1}{2}\underline{m}(\dot{\underline{c}}/\underline{c})^2 = \frac{1}{2}\underline{m}(0/\underline{c})^2 = 0$$

(D.76)

$$\dot{\underline{v}} \to \dot{\underline{c}}$$

And:

$$\text{Lim} \frac{1}{2}\underline{m}(\dot{v}/\underline{c})^2 = \frac{1}{2}\underline{m}(\dot{c}/\underline{c})^2 = \frac{1}{2}\underline{m}(0/\underline{c})^2 = 0$$

(D.77)

$$\dot{\underline{v}} \to \dot{\underline{c}} = 0$$

And:

$$\text{Lim} \frac{1}{2}\underline{V}(\dot{\underline{v}}/\underline{c})^2 = \frac{1}{2}\underline{V}(\dot{\underline{c}}/\underline{c})^2 = \frac{1}{2}\underline{V}(0/\underline{c})^2 = 0$$

(D.78)

$$\dot{\underline{v}} \to \dot{\underline{c}} = 0$$

And:

$$\text{Lim} \frac{1}{2}\underline{V}(\dot{v}/\underline{c})^2 = \frac{1}{2}\underline{V}(\dot{c}/\underline{c})^2 = \frac{1}{2}\underline{V}(0/\underline{c})^2 = 0$$

(D.79)

$$\dot{v} \to \dot{c} = 0$$

The formula $(\frac{1}{2}\underline{m}(\underline{v}/\underline{c})^2)(1-(\underline{u}/\underline{c})^2)^{+\frac{1}{2}}$, of (D.64), is the new limited temporal-scalar-speed-mass kinetic energy displacement (of increase/of decrease), for a mass $\underline{m}$.

The formula $(\frac{1}{2}\underline{m}(v/\underline{c})^2(1-(\underline{u}/\underline{c})^2)^{+\frac{1}{2}}$, of (D.64), is the new limited temporal-vector-velocity-mass kinetic energy displacement (of increase/of decrease), for a mass $\underline{m}$.

The formula $(\frac{1}{2}\underline{m}(\underline{c}/\underline{c})^2)(1-(\underline{c}/\underline{c})^2)^{+\frac{1}{2}}$ of (D.64), is the new limit temporal-vector-velocity-mass kinetic energy displacement (of zero increase/of zero decrease), for a non-existing mass m.

The formula $(\frac{1}{2}\underline{V}(\underline{v}/\underline{c})^2)/(1-(\underline{u}/\underline{c})^2)^{+\frac{1}{2}}$, of (D.65), is the new limited temporal-scalar-speed-volume kinetic energy displacement (of decrease/of increase), for the volume of a mass $\underline{m}$.

The formula $(\frac{1}{2}\underline{V}(v/\underline{c})^2)/(1-(\underline{u}/\underline{c})^2)^{+\frac{1}{2}}$, of (D.65), is the new limited temporal-vector velocity-volume kinetic energy displacement (of decrease/of increase), for the volume of a mass $\underline{m}$.

The formula $(\frac{1}{2}\underline{V}(\underline{c}/\underline{c})^2)/(1-(\underline{c}/\underline{c})^2)^{+\frac{1}{2}}$, of (D.65), is the new limit temporal-scalar speed-volume kinetic energy displacement (of zero decrease/of zero increase), for the volume of an non-existing mass $\underline{m}$.

The formula $(\frac{1}{2}\underline{m}(c/\underline{c})^2)/(1-(\underline{c}/\underline{c})^2)^{+\frac{1}{2}}$, of (D.65), is the new limit temporal-vector-velocity-volume kinetic energy displacement (of zero decrease/of zero increase), for the volume of an non-existing mass $\underline{m}$.

The formula $(\frac{1}{2}\underline{d}(\underline{v}/\underline{c})^2)/(1-(\underline{u}/\underline{c})^2)^{+\frac{1}{2}}$, of (D.66), is the new limited temporal-scalar-speed-density kinetic energy displacement (of increase/of decrease), for the density of a mass $\underline{m}$.

The formula $(\frac{1}{2}\underline{d}(v/\underline{c})^2)(1-(\underline{u}/\underline{c})^2)^{+\frac{1}{2}}$, of (D.66), is the new limited temporal-vector-velocity-density kinetic energy displacement (of increase/of decrease), for the density of a mass $\underline{m}$.

The formula $(\frac{1}{2}\underline{d}(\underline{c}/\underline{c})^2)(1-(\underline{c}/\underline{c})^2)^{+\frac{1}{2}}$, of (D.66), is the new limit temporal-scalar-speed-density kinetic energy displacement (of zero increase/of zero decrease), for the density of an non-existing mass $\underline{m}$.

The formula $(\frac{1}{2}\underline{d}(c/\underline{c})^2)(1-(\underline{c}/\underline{c})^2)^{+\frac{1}{2}}$, of (D.66), is the new limit temporal-vector-velocity-density kinetic energy displacement (of zero increase/of zero decrease), for the density of an non-existing mass $\underline{m}$.

The formula $(\frac{1}{2}\underline{m}(\dot{v}/\underline{c})^2)(1-(\dot{u}/\underline{c})^2)^{+\frac{1}{2}}$, of (D.67), is the first derivative of the new limited temporal-scalar-mass acceleration displacement (of increase/of decrease), for a mass m.

The formula $(\frac{1}{2}\underline{m}(\dot{v}/\underline{c})^2)(1-(\dot{u}/\underline{c})^2)^{+\frac{1}{2}}$, of (D.67), is the first derivative of the new limited temporal-vector-mass-acceleration kinetic energy displacement (of increase/of decrease), for a mass m.

The formula $(\frac{1}{2}\underline{m}(\dot{c}/\underline{c})^2)(1-(\dot{c}/\underline{c})^2)^{+\frac{1}{2}}$, of (D.67), is the first derivative of the new limit temporal-scalar-mass-acceleration kinetic energy displacement (of zero increase/of zero decrease), for a non-existing mass m.

The formula $(\frac{1}{2}\underline{m}(\dot{c}/\underline{c})^2)(1-(\dot{c}/\underline{c})^2)^{+\frac{1}{2}}\,\underline{m}$ if (D.67), is the first derivative of the new limit temporal-vector-mass-acceleration kinetic energy displacement (of zero increase/of zero decrease), for a non-existing mass m.

The formula $(\frac{1}{2}\underline{V}(\dot{v}/\underline{c})^2)/(1-(\dot{u}/\underline{c})^2)^{+\frac{1}{2}}$, of (D.68), is the first derivative of the new limited temporal-scalar-volume-acceleration kinetic energy displacement (of decrease/of increase), for the volume of a mass $\underline{m}$.

The formula $(\frac{1}{2}\underline{V}(\dot{v}/\underline{c})^2)/(1-(\dot{u}/\underline{c})^2)^{+\frac{1}{2}}$, of (D.68), is the first derivative of the new limited temporal-vector-volume acceleration kinetic energy displacement (of decrease/of increase), for the volume of a mass $\underline{m}$.

The formula $\frac{1}{2}\underline{V}(\dot{c}/\underline{c})^2)/(1-(\dot{c}/\underline{c})^2)^{+\frac{1}{2}}$, of (D.68), is the first derivative of the new limit temporal-scalar-volume acceleration kinetic energy displacement (of zero decrease/of zero increase), for the volume of an non-existing mass $\underline{m}$.

The formula $\frac{1}{2}\underline{V}(\dot{c}/\underline{c})^2)/(1-(\dot{\underline{c}}/\underline{c})^2)^{+\frac{1}{2}}$, of (D.68), is the first derivative of the new limit temporal-vector-volume-acceleration kinetic energy displacement (of zero decrease/of zero increase), for the volume of an non-existing mass $\underline{m}$.

The formula $\frac{1}{2}\underline{d}(\dot{v}/\underline{c})^2)(1-(\dot{\underline{u}}/\underline{c})^2)^{+\frac{1}{2}}$, of (D.69), is the first derivative of the new limited temporal-scalar-density-acceleration kinetic energy displacement (of increase/of decrease), for the density of a mass $\underline{m}$.

The formula $\frac{1}{2}\underline{d}(\dot{v}/\underline{c})^2)(1-(\dot{\underline{u}}/\underline{c})^2)^{+\frac{1}{2}}$, of (D.69), is the first derivative of the new limited temporal-vector-density-acceleration kinetic energy displacement (of increase/of decrease), for the density of a mass $\underline{m}$.

The formula $\frac{1}{2}\underline{d}(\dot{c}/\underline{c})^2)(1-(\underline{c}/\underline{c})^2)^{+\frac{1}{2}}$, of (D.69), is the first derivative of the new limit temporal-scalar-density-acceleration kinetic energy displacement (of zero increase/of zero decrease), for the density of an non-existing mass $\underline{m}$.

The formula $\frac{1}{2}\underline{d}(\dot{c}/\underline{c})^2)(1-(\dot{\underline{c}}/\underline{c})^2)^{+\frac{1}{2}}$, of (D.69), is the first derivative of the new limit temporal-vector-density-acceleration kinetic energy displacement (of zero increase/of zero decrease), for the density of an non-existing mass $\underline{m}$.[10]

Placing the above formulas into limit notation, we have:

$$(\frac{1}{2}\underline{m}(\underline{v}/\underline{c})^2)(1-(\underline{u}/\underline{c})^2)^{+\frac{1}{2}} = (\frac{1}{2}\underline{m}(\underline{c}/\underline{c})^2)(1-(\underline{c}/\underline{c})^2)^{+\frac{1}{2}} = 0$$

$$\underline{v} \rightarrow \underline{c}$$

$$\underline{u} \rightarrow \underline{c}$$

(D.80)

And:

$$(\frac{1}{2}\underline{m}(v/\underline{c})^2)(1-(\underline{u}/\underline{c})^2)^{+\frac{1}{2}} = (\frac{1}{2}\underline{m}(c/\underline{c})^2)(1-(\underline{c}/\underline{c})^2)^{+\frac{1}{2}} = 0$$

$$v \rightarrow \underline{c}$$

$$\underline{u} \rightarrow \underline{c}$$

(D.81)

And:

$$(\frac{1}{2}\underline{V}(\underline{v}/\underline{c})^2 / (1-(\underline{u}/c)^2)^{+\frac{1}{2}} = (\frac{1}{2}\underline{V}(\underline{c}/\underline{c})^2) / (1-(\underline{c}/\underline{c})^2)^{+\frac{1}{2}} = 0$$

$$\underline{v} \rightarrow \underline{c}$$

$$\underline{u} \rightarrow \underline{c}$$

(D.82)

And:

$$(\frac{1}{2}\underline{V}(v/\underline{c})^2) / (1-(\underline{u}/\underline{c})^2)^{+\frac{1}{2}} = (\frac{1}{2}\underline{V}(c/\underline{c})^2) / (1-(\underline{c}/\underline{c})^2)^{+\frac{1}{2}} = 0$$

$$v \rightarrow c$$

$$\underline{u} \rightarrow \underline{c}$$

(D.83)

And:

$$(\frac{1}{2}\underline{d}(\underline{v}/\underline{c})^2)(1-(\underline{u}/\underline{c})^2)^{+\frac{1}{2}} = \frac{1}{2}\underline{d}(\underline{c}/\underline{c})^2)(+1-(\underline{c}/\underline{c})^2)^{+\frac{1}{2}} = 0$$

$$\underline{v} \rightarrow \underline{c}$$

$$\underline{u} \rightarrow \underline{c}$$

(D.84)

And:

$$(\frac{1}{2}\underline{d}(v/\underline{c})^2(1-(\underline{u}/\underline{c})^2)^{+\frac{1}{2}} = (\frac{1}{2}\underline{d}(c/\underline{c})^2(1-(\underline{c}/\underline{c})^2)^{+\frac{1}{2}} = 0$$

$$v \rightarrow c$$

$$\underline{u} \rightarrow \underline{c}$$

(D.85)

And:

$$(\frac{1}{2}\underline{m}(\dot{v}/\underline{c})^2(1-(\underline{\dot{u}}/\underline{c})^2)^{+\frac{1}{2}} = (\frac{1}{2}\underline{m}(\dot{c}/\underline{c})^2(1-(\underline{\dot{c}}/\underline{c})^2)^{+\frac{1}{2}} = 0$$

$$\dot{v} \rightarrow \dot{c}$$

$$\underline{\dot{u}} \rightarrow \underline{\dot{c}}$$

(D.86)

And:

$$(\frac{1}{2}\underline{m}(\dot{v}/\underline{c})^2(1-(\underline{\dot{u}}/\underline{c})^2)^{+\frac{1}{2}} = (\frac{1}{2}\underline{m}(\dot{c}/\underline{c})^2(1-(\underline{\dot{c}}/\underline{c})^2)^{+\frac{1}{2}} = 0$$

$$\dot{v} \rightarrow \dot{c} = 0$$

$$\underline{\dot{u}} \rightarrow \underline{\dot{c}} = 0$$

(D.87)

And:

$$(\frac{1}{2}\underline{V}(\dot{v}/\underline{c})^2/(1-(\underline{\dot{u}}/\underline{c})^2)^{+\frac{1}{2}} = (\frac{1}{2}\underline{V}(\dot{c}/\underline{c})/(1-(\underline{\dot{c}}/\underline{c})^2)^{+\frac{1}{2}} = 0$$

$$\dot{v} \rightarrow \dot{c} = 0$$

$$\underline{\dot{u}} \rightarrow \underline{\dot{c}} = 0$$

(D.88)

And:

$$(\frac{1}{2}\underline{V}(\dot{v}/\underline{c})^2/(1-(\underline{\dot{u}}/\underline{c})^2)^{+\frac{1}{2}} = (\frac{1}{2}\underline{V}(\underline{\dot{c}}/\underline{c})/(1-(\underline{\dot{c}}/\underline{c})^2)^{+\frac{1}{2}} = 0$$

$$\dot{v} \rightarrow \underline{\dot{c}} = 0$$

$$\underline{\dot{u}} \rightarrow \underline{\dot{c}} = 0 \tag{D.89}$$

And:

$$(\frac{1}{2}\underline{d}(\dot{v}/\underline{c})^2(1-(\underline{\dot{u}}/\underline{c})^2)^{+\frac{1}{2}} = (\frac{1}{2}\underline{d}(\underline{\dot{c}}/\underline{c})^2(1-(\underline{\dot{c}}/\underline{c})^2)^{+\frac{1}{2}} = 0$$

$$\dot{v} \rightarrow \underline{\dot{c}} = 0$$

$$\underline{\dot{u}} \rightarrow \underline{\dot{c}} = 0 \tag{D.90}$$

And:

$$(\frac{1}{2}\underline{d}(\dot{v}/\underline{c})^2(1-(\underline{\dot{u}}/\underline{c})^2)^{+\frac{1}{2}} = (\frac{1}{2}\underline{d}(\underline{\dot{c}}/\underline{c})^2(1-(\underline{\dot{c}}/\underline{c})^2)^{+\frac{1}{2}} = 0$$

$$\dot{v} \rightarrow \underline{\dot{c}} = 0$$

$$\underline{\dot{u}} \rightarrow \underline{\dot{c}} = 0 \tag{D.91}$$

Notes

¹ This is the first time in the History of Physics that we have defined the temporal-motive displacements (or of the time-motive displacements), of a mass m, having the speed +v, or the vector velocity +v. We now have the mass term +m, having the speed term: +v, or having the vector velocity term: +v. And which is divided by the speed limit +c of light. We then have: +m(v / c), or: +m(v / c). We do not measure the temporal-motive displacements (or the time-motive displacements), *by the distance a mass* +m *will travel.* Such as, it takes a year for the Earth travelling around the sun. But instead, if any moving mass has the speed +v, or a vector velocity +v. Hence, by dividing these two limited motion terms by the speed +c of light, which gives us a new conception of temporal-motive displacements; or of time-motive displacements, that a mass +m will display. We also have the limit temporal-motive displacements; and while the paragraph above is the limited temporal-motive displacements. And whereas, we then will use the limit scalar speed term: +c, of light. And we have (+c / c) = +1, of one limit scalar second. Or dividing the limit vector velocity term: +c, into itself, we have (+c / c) = +1, of one vector second. And now when we multiply the mass term: +m, into these expressions, a new limit temporal-motive displacements, or a limit time-motive displacements, of one scalar second, or of one vector second.

² We must note in regards to the units of measurements, one temporal-motive kilometer displacement is equal with one motive kilometer displacement.

³ Although, in this Appendix D, we will give many formulas on this new concept of "temporal-motive" displacements. this is because these many formulas are necessary in order to advance this new concept of "temporal-motive displacements."

⁴ When we use this new vector second term: +1, of one limit vector second, this one limit vector second has vector direction, and a magnitude of one scalar second +1. Hence, we can discuss vector temporal-motive displacements as having uniform vector direction displacement.

⁵ It is more precise to consider the limit vector velocity term: +ċ to be a zero value, because the first derivative of a constant +c to be a zero value "0"; instead of using the limited vector term: +v̇ As being a zero value "0".

6   More precisely the first derivative of the limit vector velocity term: $+c$ is ($\dot{+} = 0$), as well as the first derivative of $+1$ is ($\dot{1} = 0$), and where $+\dot{v} > 0$.

7   In this section—e-, we will deal with the kinetic-energy formulas of volume-motive displacements, and the kinetic energy formulas of density-motive displacements. As well as the kinetic energy formulas of mass-motive displacements.

8   Albert Einstein in his physic's paper proposed the following relativistic formula: $(+L_0) / (+1 - (\underline{u} / \underline{c})^2)^{+\frac{1}{2}}$, for the length contraction of body in motion, and which it approaches the speed limit $+\underline{c}$ of light in vacuum.

Unfortunately, Albert Einstein, and the many proponents of his special relativity theory, concluded that a bar of mass, or steel, would contract in the direction of the moving mass, or the moving bar of steel.

I also reject Albert Einstein's length contract formula for another reason. Since, Albert Einstein, and the many proponents of his special relativity theory treat the limiting factor: $(+1 - (\underline{u} / \underline{c})^2)^{+\frac{1}{2}}$ as some kind of force-factor And which this relativistic factor somehow changes the $+\underline{L}_0$ into another length $+\underline{L}$, where $+L < +L_0$ But this limiting factor does not do this.

In fact all of Albert Einstein's relativistic formulas are false measuring formulas, and it should be recognized as such, as based upon our new mechanical theory. Many proponents of Albert Einstein's conception of contraction of a bar of mass, or of steel—within a box1 travelling throughout space and time. And then another box2, the observers of box1 could see the identical bar of mass, or of steel, would display this so called length contraction. And while, the observers of box1 could see the length contraction of the bar of mass, or of a steel bar, of box2, would see the bar contract within the box2. And neither of these two moving boxes 1 and 2, could not detect their own bar of mass, or of steel, would see any contraction taken place.

8   Instead of the relativistic length contraction formula, I intend, throughout this main text of this book, as well as in Appendix D, to use instead a volume term: $+\underline{V}$, in the temporal-motive formulas, as well as in the main text of this book.

9   As we have not given formula (D.27) in the text of this book, we will give the formula:

$$\underline{m} / \underline{V}(\underline{c} / \underline{c}) = \underline{d}(\underline{c} / \underline{c}) = \underline{d}(\underline{1})$$

And quoting: "the division of the volume $\underline{V}$ into mass $+\underline{m}$, as based upon formula (D.27), results in density. Where the terms: ($\underline{c}$ / $\underline{c}$) are of a limit, yet a constant, a temporal-motive density-momentum displaced formula, in regards to its density, and its mass". End quote.

[10] The limit notations in this Appendix D, are self-evident and do not need any further interpretations.

The terms $+\underline{1}$, and $+\underline{v}$, etc., are scalar terms, and while the terms $+1$, $+v$, etc., without any under lines are vector terms.

APPENDIX E

A NEW MODEL FOR GRAVITATING MASSES

-a-

Around the time of January 17, 2006, I made a thought experiment involving two equal yet small steel spherical masses, where: $+\underline{m}_1$, and $+\underline{m}_2$, and where $+\underline{m}_1 = +\underline{m}_2$. And also involved with this thought experiment, I had used two equal applied forces: F_1 and F_2, and where: $F_1 = F_2$. And whereas, these two equal spherical masses: $+\underline{m}_1$, and $+\underline{m}_2$, are within the gravitational field of the Earth. And these two equal applied forces: F_1, and F_2 also operates within the same gravitational field of the Earth.

In my thought experiment I had constructed smooth planks of wood, the width of the planks are two feet across. And the length of the planks of wood are such as 100 feet of length. And which these planks of wood are layed flat down on a smooth and level ground of the Earth's ground. And these planks of wood are connected together with each other by glue, or some other kind of fasteners of these planks of wood. And these planks of wood can have sides of additional wood about one half of a foot on all the sides of these planks of wood.

We now have, according to my thought experiment, two equal forces: F_1 and F_2, are to be used as explosive devices. And one force: F_1 is layed beneath, or under the mass $+\underline{m}_1$ on a level ground. And when this one force: F_1 detonates beneath, or under the mass $+\underline{m}_1$, within the Earth's gravitational field. And thus, the mass $+\underline{m}_1$ will assume a vertical height of $+\underline{h}_1$, and the same mass $+\underline{m}_1$ will assume a vertical height of $+\underline{h}_1$, and the same mass $+\underline{m}_1$ will free fall from the same height of $+\underline{h}_1$. And the total distance the mass $+\underline{m}_1$ will travel is: $+2\underline{h}_1$.

According to the second mass $+\underline{m}_2$, and the second force: F_2, where $+\underline{m}_1 = +\underline{m}_2$, and $F_1 = F_2$. And the spherical mass $+\underline{m}_2$ is made of steel, and it is placed upon the beginning of the plank's wood length, in which these planks of wood are very smooth.

And thus, when another explosive device detonates (the dynamic action of the force: F_2), at ground level, upon the mass $+\underline{m}_2$, within the Earth's gravitational field. Then the mass $+\underline{m}_2$ will acquire a uniform motion, and will move a distance of length $+L_1$, and thus: $+\underline{L}_2 > +2\underline{h}_1$. That is, the mass $+\underline{m}_2$, on the horizontal planks of smooth wood, will travel, horizontally a greater distance that the mass $+\underline{m}_1$, which traveled, up and down, vertically, the distance $+2\underline{h}_1$. And these events are happening within the Earth's gravitational field. These thought experiments can be carried out by experimental physicists, which will establish the truthfulness of my thought experiment.

Where the two masses $+\underline{m}_1$ and $+\underline{m}_2$ have vertical and horizontal movements, within the Earth's gravitational field. And where these two uniform motions are caused by the dynamic actions of two equal forces: F_1 and F_2, and also $+\underline{m}_1 = +\underline{m}_2$. And thus $+\underline{L}_1$; $+2\underline{h}_1$ in distances traveled by these two equal masses. We now conclude that $+\underline{L}_1 > +2\underline{h}_1$. Or the mass $\underline{m}_2$ travels a greater distance, horizontally, than the mass $+\underline{m}_1$, which travels a less distance, vertically. And both masses are moving within the Earth's gravitational field.

-b-

According to a electro-magnetic force, when a magnet is "pushed" across a metal sheet, the force used to "push" the magnet, a small distance, is much less in force, than the force used to raise vertically, the same magnet, from the metal sheet.

And I conjecture that these same models may be used for the "weak force". And for the "strong force"—in regards to their horizontal movements, and in comparison to their vertical movements. And which it takes less force to move the "weak force" particle horizontally, than its vertical movement. And it takes less force to move the "strong force" particle horizontally, than its vertical movement.

-c-

To explain this thought experiment, I considered that the horizontal motion of the steel ball travels a path of least gravitational actions, in regards to the Earth's gravitational field. And I also considered that the vertical movement of the other steel ball travels a path of greater gravitational actions, in regards to the Earth's gravitational field.

And in regards to all the planets revolving around the sun, the planet's orbits can be considered, empirically, as *being horizontal elliptic orbits*. And which these elliptic orbits of the planets *travels paths of least gravitational actions*, with respect to the sun's gravitational field.

One explanations of these facts of the horizontal movement of a ball, and the vertical movement of another ball, I proposed a new model. Which is: that the gravitational field of the Earth *increases with distance, and decreases with nearness*. Hence, this model explains why the ball on a horizontal plane travels a greater distance that the vertical ball, which travels a less distance. But I reject this model, as being too complicated and it implies strange physical effects.[1] Therefore, I accept the traditional view that the gravitational field is a weak field[2], which becomes weaker with distance.[3]

Notes

[1] I now reject this later model that the gravitational field increases with distance, and it decreases with nearness between two gravitating masses: $+m\dot{v}$, and $+m\dot{v}$.

[2] Sometimes the simple experiment can result in large empirical results.

[3] Gravitation seems to be a strong vertical force-field in its strength—while gravitational horizontal force-field in its strength is not-strong.

APPENDIX F

THE SUB UNIVERSES

-a-

Our entire universe is made of seven, or more sub universes. And the two sub universes that we will use in this Appendix F, are the: *great material sub universe*, and the *great electro-magnetic radiations sub universe*. The "great material sub universe" is composed of all the mass in the entire universe. And the "great electro-magnetic radiations sub universe" is composed of the entire spectrum of all the electro-magnetic radiations phenomena, in the entire universe.

There are five more sub universes. The third one is the *great neutrio sub universe*. It has been said that a trillion neutrios' pass throughout our bodies every second. The fourth sub universe is the *great dark matter, and dark energy sub universe.*[1] The fifth sub universe is the *great gravitational force-fields sub universe*. And the six sub universe is the *great electro-magnetic force-fields sub universe*. And the seven sub universe is the *great strong and weak force-fields sub universe.*

All of these seven sub universes[2] makes our entire universe, and in this Appendix F, we will use the two sub universes of: the "great material sub universe", and the "great electro-magnetic radiations phenomena sub universe".

-b-

The terms: $+\underline{c}$, and $+\underline{1}$, are with respect to all the sub-universes of the "great electro-magnetic radiations phenomena sub universe". Which has the common speed limit $+\underline{c}$ of light, of one second, in free vacuum space and time. And as measured in units of $3(10^4)$ kilometers per one second.

To place these two terms into limit notations, we have:

$$\text{Lim} + \underline{v} = +\underline{c}$$
$$+\underline{v} \rightarrow +\underline{c}$$

$$\text{Lim} + \underline{\dot{v}} = +\underline{\dot{c}} = 0$$
$$+\underline{\dot{v}} \rightarrow +\underline{\dot{c}} = 0$$

$$\text{Lim}\, v = +c$$
$$+v \rightarrow +c$$

$$\text{Lim}\, \dot{v} = +\dot{c} = 0$$
$$+\dot{v} \rightarrow +\dot{c} = 0$$

On the left hand side of these limits are for the speed, and vector velocity of a mass $+\underline{m}$, in free vacuum—with respect to the sub universe of the: "great material sub universe". And on the right hand sides of these limits are of the: "great electro-magnetic radiations phenomena sub universe". And dividing these limits by the terms: $+\underline{c}$, c, $+\dot{c}$, and $+\underline{\dot{c}}$, we have:

$$\text{Lim} + \underline{v} / +\underline{c} = +\underline{c} / +\underline{c} = +\underline{1}$$
$$+\underline{v} \rightarrow +\underline{c}$$

$$\text{Lim} + \underline{\dot{v}} / +\underline{\dot{c}} = +\underline{\dot{c}} / +\underline{\dot{c}} = +0$$
$$+\underline{\dot{v}} \rightarrow +\underline{\dot{c}} = 0$$

$$\text{Lim} + v / +\underline{c} = +c / +\underline{c} = +1$$
$$+v \rightarrow +c$$

$$\text{Lim} + \dot{v} / +\underline{c} = +\dot{c} / +\underline{c} = +0$$
$$+\dot{v} \rightarrow +\dot{c} = 0$$

And we have:

$$\text{Lim} + \underline{v} / +c = +c / +c = + 1$$
$$+\underline{v} \rightarrow +\underline{c}$$

$$\text{Lim} + \underline{\dot{v}} / +\dot{\underline{c}} = +\dot{\underline{c}} / +\dot{\underline{c}} = +0$$
$$+\underline{\dot{v}} \rightarrow +\dot{\underline{c}} = 0$$

$$\text{Lim} + v/ + c = +c/ + c = +1$$
$$+v \rightarrow +c$$

$$\text{Lim} + \dot{v} / +\dot{c} = +\dot{c} / +\dot{c} = +0$$
$$+\dot{v} \rightarrow +\dot{c} = 0$$

And we conclude that the sub universe of the "great material sub universe"—in regards to all kinds of mass of the universe has "limited time", and "limited motion". As on the left hand side of these limit notations of this page indicates, and as the left hand side of the previous page indicates.[3] And while on the right hand side of these limit notations has "limit time", and "limit motion", for all the "great electro-magnetic radiations phenomena sub universe".

And because of our use of the concept "limit", we have "limited time", as a variable, which as a limit of one scalar second: $+\underline{1}$. Such as:

$$\text{Lim}(+\underline{v} / +\underline{c}) = +\underline{1}$$
$$+\underline{v} \rightarrow +\underline{c}$$

And we also have:

$$\text{Lim} + \underline{v} = +\underline{c}$$
$$+\underline{v} \rightarrow +\underline{c}$$

Where the term: $+\underline{c}$, is the scalar speed limit, a constant, of light, throughout the entire universe. And the term: $+c$ is called the "limit speed displacement". And the term: $+\underline{v}$ is called the "limited speed displacement". And the term: $+\underline{1}$ is called the "limit temporal-speed displacement". And the term: $+\underline{v} / +\underline{c}$, is called the: "limited temporal-speed displacement".

-c-

Throughout the entire universe, there exists the total sub universe of the "great electro-magnetic radiations phenomena sub universe". Which is to say, as far as we may travel in the entire universe—which is not very far—we would experience the mind expansion of the entire universe in its wonderful "great electro-magnetic radiations phenomena sub universe".

Hypothetically, we infer, as based upon our experiences of the entire universe—which is not much experience—that the "great electro-magnetic radiations phenomena sub universe", has existence because of the existence of the entire universe. There are not any locations in the entire universe, except for the phenomena of "black holes", which the "great electro-magnetic radiations phenomena sub universe" would not exist for us.

And we conclude, as based upon the above results, that all "great electro-magnetic radiations phenomena sub universe" has: when this sub universe, *taken as a whole, or as an entirely, has the universal and constant speed limit +$\underline{c}$ of light, in free vacuum space and time*. And which has the measuring units of $3(10^4)$ kilometers *per one second*.

It is most appropriate, as based upon the above results, that by using the term: +$\underline{c}$, as a scalar speed limit of light, in free vacuum, *that we will derive a new and universal constant of Nature*. And this new constant is of one scalar second: +$\underline{1}$. And this new constant of Nature *is derived by dividing the speed limit +$\underline{c}$ into itself*. And which this speed limit +$\underline{c}$ of light, measures the speed limit, an universal constant, of the "great electro-magnetic radiations phenomena sub universe". Hence, we have a new constant of one scalar second +$\underline{1}$.

Thus, we have made a bold proposal that the time unit to measure the uniform speed limit +$\underline{c}$ of the "great electro-magnetic radiations phenomena sub universe"; throughout the entire universe is: this new time unit, a universal and constant time unit, of one scalar second: +$\underline{1}$. And we consider this new time unit of one scalar second: +$\underline{1}$, in its own right to be placed as being a new derived universal constant of Nature.[4] And this new time unit of one scalar second: $+\underline{1}$, is a new derived time limit for all the limited times, within our entire universe.

All astronomers know, and perhaps all laypeople know, when they observe the night sky: "of the great electro-magnetic radiations phenomena sub universe"; (or observe parts of it). They then know of what they observe is the "cosmic, or universal past-time". They then conclude for all intelligent life on the Earth, with respect to the "great electro-magnetic radiations phenomena sub universe", *is that they live in a universal past time*, with respect to the entire universe.

I disagree with this conclusion. Since, as we have seen, the "great electro-magnetic radiations phenomena sub universe" has—throughout the entire universe,—the common and constant speed limit $+\underline{c}$ of light, a constant, in free vacuum space and time. That is, when this "great electro-magnetic radiations phenomena sub universe" is taken individually, or taken as a whole—(taken in its entirely). And also when this same sub universe has a new and derived and universal temporal-speed constant of one limit temporal-second: $+\underline{1}$—with respect to all the other constants of Nature. This temporal-speed constant of one scalar second: $+\underline{1}$, serves as a "limit" for all the other "limited great material sub universe". And which the "limited great material sub universe" have the limited temporal-motive displacements of: $+\underline{v}/+\underline{c}$, $+\underline{\dot{v}}/+\underline{c}$, $+\underline{\dot{v}}/+\underline{c}$, $+v/+\underline{c}$, $+\underline{v}/+c$, $+v/+c$ and: $+\underline{\dot{v}}/+\underline{\dot{c}}$, $+\underline{\dot{v}}/+c$, $+\underline{\dot{v}}/+\dot{c}$, $+\dot{v}/+\dot{c}$.

And all of these fractional variables has as there limit of one scalar second: $+\underline{1}$, or one vector second: $+1$. And all of these "limited terms" express the "limited temporal-motive displacements" for all "limited great material sub universe" taken individually, or taken as a whole—with respect to the "great limited material sub universe".[5]

Proponents of Albert Einstein's special relativity theory measures the entire universe in sequences of larger increasing numbers, such as: $(1,2,3, \ldots ,\underline{n})$ in "past light seconds", in "past light minutes", "in past light months", or "in past light years". Hence, according to these proponents, we live in "past light times", ie., we live in the cosmic past with respect to our entire universe.

But we measure the sub universe of the "great electro-magnetic radiation phenomena sub universe" in regards to a unlimited sequences of natural constants. Whereas, the natural sequence of the constant limit +c̲ of light in free vacuum. And also of the new sequence of the universal and limit constant of the scalar one second: +1̲. Or of the vector constant of one vector second: +1. And which these two constants are used to measure the speed constant and limit of light +c̲ for the "great electro-magnetic radiation phenomena sub universe".

We now have: an unlimited sequence of: (+c̲, +c̲, +c̲, ... +c̲), which is to be applied to the "great electro-magnetic radiation phenomena sub universe" in regards to its fundamental constant and speed limit +c̲ of light, as it is measured by one second, in free vacuum space and time. As the speed limit +c̲ can be measured in a sequence of one scalar second: (1̲, +1̲, +1̲, ... +1̲). Or in a sequence of one vector second: (+1, +1, +1, ... +1). And these one seconds are derived constants in comparison to the measured constant speed limit +c of light, as measured by one second, in free vacuum.

These sequences of natural constant means, empirically, that we observe, (or infer), the "great electro-magnetic radiation phenomena sub universe—as it is happening at the sequence (+c̲, +c̲, +c̲, ... , +c̲) at the speed limit +c̲, of light, in free vacuum space and time. We also have, in the measuring of these speed limit's +c̲, a new derived natural constant, and a limit, which is the natural sequence of an unlimited scalar one second: (+1̲, +1̲, +1̲, ... , +1̲). Or of the sequence of one vector second: (+1, +1, +1, ... +1). And which these constants and limits measures the "great electro-magnetic radiation phenomena sub universe", and also measures the "great limited material sub universe".

In contrary of what some astronomers, and physicists, may say, we must conclude that we live in a "great electro-magnetic radiation phenomena sub universe". And which can be called the *Universal Now Time*, or the *Limit Universal Constant Now Time*—of the unlimited sequence of one scalar second. That is, when: (+1̲, +1̲, +1̲, ... , +1̲). Or of a vector sequence: (+1, +1, +1, ... +1).

We also live within the *Limit Universal Now-Speed Displacement*, with respect to a sequence of: (+c̲, +c̲, +c̲, ... , +c̲), of the speed limit and constant +c̲ of light, in free vacuum space and time. Or we also live within the *Limit Vector-Velocity Now Displacement*, with respect to a sequence of: (+c, +c, +c, . . . , +c), of the vector-velocity limit and constant +c, of light, in free vacuum space and time. And which the terms: +$\underline{c}$'s, and +c's, measures all of the "great electro-magnetic radiation phenomena sub universe", throughout the entire universe.

Most, if not all, astronomers and physicists conclude that our universe is of a "great cosmic past time universe". As these astronomers and physicists measure the "great electro-magnetic radiation phenomena sub universe". This is the case, for all Earth bound sky observers upon the Earth. These same scientists use a sequence of increasing numbers: (+1, +2, +3, . . . , +n) to measure the "light seconds", the: "light minutes", and the "light months", and the "light years" to measure the light distance between the sources of the "great electro-magnetic radiation phenomena sub universe", throughout the entire universe.

Why do we live in a "cosmic part time universe?" It is because the phenomena of light has a finite speed limit +$\underline{c}$, measured in units of $3(10^4)$ kilometers per one second. And because of the tremendous distances (in light years distances), between all the stars and galaxies, which we observe and infer from our planet Earth.

Unfortunately, most scientists do not inform the public, who may be interested in the universe, or not, that all people kind live with a "cosmic past time universe". When we look at the stars, at night, we are only seeing the "past" of these "stars" and "galaxies". And our observations of the "past" of the entire universe, is caused by the finite speed limit +$\underline{c}$ of light, as measured by one second, in free vacuum space and time. Our observation of the "past" when we view the night sky, is due to the great light distances between the stars, and between the galaxies, in regards to our entire universe.

-d-

As we have considered the "limit now-time", in regards to the "now-time of light-speed", or in regards to the "now-time of light-vector-velocity"—with respect to the "speed" +c̠ of light, or with respect to the "vector-velocity" +c, of light, in free vacuum space and time.

In this section—d-, I will give the "limited variable time", and also I will deal with the "limited variable speed". This concept of "limited" deals successfully with the "great limited material sub universe".

We can put the "limited variable time"; and put the "limited variable speed"—into limit notations (see the appendices, and section—c-, of Appendix F).

We now have:

$$\text{Lim} + \underline{v} = +\underline{c}$$
$$+\underline{v} \rightarrow +\underline{c}$$

$$\text{Lim} + v = +c$$
$$+v \rightarrow +c$$

$$\text{Lim} + \underline{\dot{v}} = +\underline{\dot{c}} = 0$$
$$+\underline{\dot{v}} \rightarrow +\underline{\dot{c}} = 0$$

$$\text{Lim} + \dot{v} = +\dot{c} = 0$$
$$+\dot{v} \rightarrow +\dot{c} = 0$$

Where the limit notations displays the "limited variable speed displacements" on the left hand side of these limit notation, is the "limit now-speed displacements". And from these limit notations, we see that the speed limit +c̠ of light is the upper limit, or/and the lower limit (0), when this term: +c̠, is multiplied into a mass term +m̠. Or that the vector-velocity limit +c is also an upper limit, or/and a lower limit (0), when this term: +c, is multiplied into a mass term: +m̠. And these limits are for all the motive-mass-displacement with respect to the "great limited material sub universe". No mass,[6] within the "great limited material sub universe", can ever travel with the speed limit and universal constant of light, in free vacuum, as light is measured by a time interval of one second.

We now have the temporal-motive notation formulas:

$$\text{Lim} + \underline{v} / +\underline{c} = +\underline{c} / +\underline{c} = +\underline{1}$$
$$+\underline{v} \rightarrow +\underline{c}$$

$$\text{Lim} + v / +\underline{c} = +c / +\underline{c} = +1$$
$$+v \rightarrow +c$$

$$\text{Lim} + \underline{\dot{v}} / +\underline{c} = +\underline{\dot{c}} / +\underline{c} = +0$$
$$+\underline{\dot{v}} \rightarrow +\underline{\dot{c}} = 0$$

$$\text{Lim} + \dot{v} / +\underline{c} = +\dot{c} / +\underline{c} = +0$$
$$+\dot{v} \rightarrow +\dot{c} = 0$$

These four new limits are temporal-motive displacement formulas.[7] The temporal-motive displacement formulas are derived for any moving mass, within the "great limited material sub universe". And any mass may have a "motion displacements". The "temporal-motive displacement" for light is one scalar second: $+\underline{1}$, or one vector second: $+1$. And these temporal-motive displacements, for any mass $+\underline{m}$, within the "great limited material sub universe", is of one scalar variable, or one vector variable, as the terms on the left hand side of these limits indicates—with respect to one limit, constant, and universal second, a scalar term: $+\underline{1}$, or a vector term: $+1$.

As an example, the mass term $+\underline{m}$'s temporal-motive displacement is of one vector, limited, and variable temporal-motive term: $+v / +\underline{c}$, with respect to one vector, limit, constant, and universal second: $+1$.

I came up with the idea or concept of "temporal-motive" when I realized that a moving light beam, or a moving mass, has a "time" phenomena connected to the moving light beam, or to the moving mass. I derived the "temporal-motive" displacement of light, by dividing the speed $+\underline{c}$ of light into itself, or dividing the vector velocity $+c$ of light also into itself, and thus: coming up with a new derived constant and time limit in its own right. And we can place this new time limits, ie., $+\underline{1}$, and $+1$, which have been derived, among all the other measured natural constants of Nature.

The universe we are experiencing is of a "now-time", of a constant, universal, and a limit now-time: of one scalar second: +$\underline{1}$, or of one vector second: +1, in an unlimited sequences of scalar seconds, or in an unlimited sequences of vector seconds. This universe is the "great electro-magnetic radiation phenomena sub universe". This is the truth when we measure all the universal light radiation phenomena by an unlimited sequence of one scalar seconds: (+$\underline{1}$, +$\underline{1}$, +$\underline{1}$, . . . , +$\underline{1}$). Or when we measure all the universal light radiation phenomena by an unlimited sequence of one vector seconds: (+1, +1, +1, . . . +1). We can inductively infer that all the "great electro-magnetic radiation phenomena sub universe" throughout the entire universe, is measured by an unlimited sequence of scalar one seconds. Or can be measured by an unlimited sequence of vector one seconds. And we can infer the bulk of the "great electro-magnetic radiation phenomena sub universe", when taken as a whole, without bring into our measurements, the: "distance factor", between stars, planets, galaxies, and other cosmic objects. Hence, we can measure the entire universe, in regards to its "great electro-magnetic radiation phenomena sub universe" by an unlimited sequence of scalar second terms: (+$\underline{1}$, +$\underline{1}$, +$\underline{1}$, . . . , +$\underline{1}$). Or by an unlimited sequence of vector second terms: (+1, +1, +1, . . . , +1).

Most, if not all, astronomers, and physicists measure the "light universe" by making an increasing unlimited sequence of unequal natural numbers, ie., (+$\underline{1}$, +$\underline{2}$, +$\underline{3}$, . . . ,, +$\underline{n}$). And these scientists conclude that we do live in a "universal past"—whenever we observe the nightly sky. And every where we go on the Earth, and observe the nightly sky it is a "ghostly reminiscent of all past life in the entire universe".

But in contrast we can measure the "light universe", to its limit, and by using the conception of "now-time", by inferring the entire "light-universe" is the same from any point in our unique universe, or our unique "great electro-magnetic radiation phenomena sub universe". Such as, that the entire "light-universe" is always measured by one sequence of an unlimited scalar seconds: (+$\underline{1}$, +$\underline{1}$, +$\underline{1}$, . . . , +$\underline{1}$). Or measured by one sequence of an unlimited vector seconds: (+1, +1, +1, . . . , +1). And since, all of these scalar seconds are the same, and also since, all of these vector seconds are the same—this then implies we can measure the entire "light-universe" by one scalar second, or by one vector second.[8]

And we also measure the limit speed $+\underline{c}$ of light, measured in $3(10^4)$ kilometers per one second. Or also measure or measured the limit vector velocity $+c$ of light where it is measured by $3(10^4)$ kilometers per one second, or one vector second. And the phenomena of light is a fundamental limit, constant, and universal, scalar or vector, for all the "great electro-magnetic radiation phenomena sub universe" within the entire universe. This is our "grand inference" that we make about the "great electro-magnetic radiation phenomena sub universe" which is the same any where within this entire universe.

We must conclude on how "strange" it may seem, that we live in a "cosmic past universe", and also live within a "cosmic now-time universe". These two kinds of sub universes depends upon how we measure the "great electro-magnetic radiation phenomena sub universe". These two kinds of sub universes depends upon the two different sequences we may use to measure these two kinds of sub universes. Such as: for the limit speed $+\underline{c}$ of light, is measured in an increasing sequence of equal seconds: $(+\underline{1}, +\underline{1}, +\underline{1}, \ldots, +\underline{1})$. Or is measured by an increasing sequence of vector seconds: $(+1, +1, +1, \ldots, +1)$. And which has the constant, and limit, and universal sequence of the speed limit $+\underline{c}$ of light in free vacuum: $(+\underline{c}, +\underline{c}, +\underline{c}, \ldots, +\underline{c})$. Or an universal sequence of the vector velocity limit $+c$ of light in free vacuum: $(+c, +c, +c, \ldots, +c)$.

We must also note that the "great limited material sub universe", taken in its entirely, as a small part, has limited motion, and limited time. And throughout this book, we have been showing that the "limited motive-mass displacements", which can never equal the speed limit $+\underline{c}$ of light, as it is measured by one second, in free vacuum. And we have discovered a new concept of "temporal-motive displacements". Whereas, all "limited times", are with respect to the motion of a mass $+\underline{m}$, and with respect to a force: $+F_v$—are said to have limited fractional variables, such as:[7] $+v / +\underline{c}$, and: $+\underline{v} / +\underline{c}$. And these two fractional variable temporal-motive displacement terms are said to be "limited". These two motion terms will never equal the limit-temporal-motive displacement terms: $+\underline{c}$ / $+\underline{c}$, and $+c$ / $+\underline{c}$, of one scalar second, or of one vector second.

Notes

[1]   I have rejected the sub universe of dark matter and energy. I am not sure whether dark matter, and dark energy really exists within our universe.

[2]   We can now list all of the various kinds of "sub universes". "The first great bulk material sub1 universe". "The second great bulk gravitational sub3 universe". "The third great electro-magnetic radiations sub3 universe". "The fourth great molecular sub4 universe". "The fifth great bulk atomic sub5 universe". "The sixth great bulk electron sub6 universe". "The great seventh bulk nucleus sub7 universe". "The great eight bulk quark sub8 universe". "The great bulk ninth weak force sub9 universe". "The great bulk tenth strong force sub10 universe". "The great bulk eleventh sub11 universe". And "the great bulk elementary twelfth particles sub12 universe".

 And by using this term of "subs" is because we can consider the total universe, on the microscopic scale, and on the macroscopic scale, to be of the "total set universe".

[3]   And by using this term, or notion of "subs", we can distinguish between the various kinds of "sub universes". And by using this empirical notion of "subs" we can reveal the connections between the various kinds of "sub universes", with respect to the "total set universe". I have forgotten to include the thirteen electro-magnetic force-fields sub13 universe. And the fourteen neutrino's sub14 universe. As well as the fifteen Background Radiation sub15 universe. As well as the sixteen spacetimes sub16 universe. As well as for other sub universes not mentioned in this note—2-.

 Why have I used this new empirical conception of "sub universes?" It is because we can deal with one sub universe at one time. And we can show how all the sub universes are connected together, within respect to the total set universe.

[4]   The "limited motive terms" are the following: $+\underline{v}$, (a limited scalar term). The term $+v$, (a limited vector term). The term $+\dot{v}$, (the first derivative of

the term: $+\underline{v}$. The term $+\dot{v}$, (the first derivative of the term: $+v$). And while the "limited temporal-motive terms" are the following: The term $+\underline{v}$ / $+\underline{c}$, (a scalar temporal-motive limited term). The term: $+v$ / $+\underline{c}$, (a limited vector temporal-motive term). The term $+\dot{v}$ / $+\underline{c}$, (a limited scalar temporal-motive first derivative of the term: $+\underline{v}$). The term: $+\dot{v}$ / $+\underline{c}$, (a limited vector temporal-motive first derivative term: $+v$).

The limit motive-terms are the following: $+\underline{c}$, the speed of light, in free vacuum. And the term: $+c$, the vector velocity of light, in free vacuum.

And the limit temporal-motive terms are the following: $+\underline{c}$ / $+\underline{c}$ = $+\underline{1}$, (which is the limit temporal-motive of one scalar second. And the term: $+c$ / $+\underline{c}$ = $+1$, (which is the limit temporal-motive of one vector second. And the limit temporal-motive term is: $+\dot{c}$ = 0, (which is the first derivative of one $+\dot{c}$ / $\underline{c}$ =0 term). And the limit temporal-motive term: $+c$ / $+\underline{c}$ =0, (which is the first derivative of one zero terms). And while the terms: $+\underline{1}$ is of one scalar second. And while the term: $+1$, is of one vector second.

And while the limit motive terms are the "limit' for all the limited motive terms. And while all the limit temporal-motive terms are the "limit" for all limited temporal-motive terms.

5   The "great material sub universe", throughout the entire universe, has "limited time", and "limited motion". And while the "great electro-magnetic radiations sub universe", throughout the entire universe, has "limit time", and "limit motion". And the "limit time" is the limit of "limited time". And the "limit motion" is the limit of "limited motion".

6   Physicists and astronomers measure the great distances of the stars, and galaxies, from the vantage point of the Earth. They use the empirical concept of: "lightyears". And a year can be reduced to an X number of seconds. Where the X number of seconds equals one year. Hence, we multiply the limit and constant speed $+\underline{c}$, of light, in free vacuum, into the X number of seconds. And thus, the formula, or term: $+\underline{c}X$ is then the distance that light had traveled in X number of seconds, which is then one lightyear.

7   I also had another new idea, as the speed limit $+c$, of light in free vacuum, as measured by one second. Hence, we have the following relation: $+c \sim +1$, which the symbol $\sim$ means that the speed limit of light, is equivalent to the

time limit of one second. And in this Appendix F, we have treated the new conception of a sequence of one identical seconds; thus: (+$\underline{1}$, +$\underline{1}$, +$\underline{1}$, . . . , +$\underline{1}$). And as based upon this time sequence we will come up with the new idea of the universal constant and limit "now time displacements", with respect to this sequence of repeating one seconds—throughout the entire electro-magnetic radiations sub universe.

This new concept of the *universal now time displacements* has originated because of the *universal constant and limit speed* +$\underline{c}$ of light, in free vacuum. And we have formulated a speed limit +$\underline{c}$ of light, in free vacuum space and time—within the following *limit now motion displacements*, as: (+c, +$\underline{c}$, +$\underline{c}$, . . . +$\underline{c}$) And this sequence of motion +$\underline{c}$, are the *universal constant and limit "now motion" displacements*; throughout the entire electro-magnetic radiations sub universe.

And when we multiply the first sequence by a number N, we would have: +N($\underline{1}$, +$\underline{1}$, +$\underline{1}$, . . . +1) = (+N, +N, +N, . . . +N). And whereas, on the right hand side of this relation is the *universal constant and limit now time* +$\underline{N}$ *displacements*. Which means that we measure the sequence of the universal constant and limit speed +N$\underline{c}$, of light, in free vacuum, as based upon the relation: +N(+$\underline{c}$, +$\underline{C}$, +$\underline{c}$, . . . +$\underline{c}$) = (+N$\underline{c}$, +N$\underline{c}$, +N$\underline{c}$, . . . +N$\underline{c}$).

7   Whereas, on the right hand side of this relation means that light has traveled a distance +N$\underline{c}$, as measured by +N(1) seconds. Hence, from our vantage point on Earth, our entire universe is of an *universal constant limit now time displacements*. And also that our entire universe is of a *universal constant and limit now speed, or motion displacement*.

Then the question needs to be ask: why are the stars, and galaxies, of our entire universe have such great distances from our vantage point of our Earth? The answer to this question is that the "entire material sub universe" is limited in time displacement, and limited in motion displacement. For all practical reasons, our total set universe is a "now universe". And when we observe the night sky we are experiencing the "now universe"; in regards to the time of this "great electro-magnetic radiations sub universe", and also in regards to the "now motion" of the "great electro-magnetic radiations sub universe". This is the truth, eventhough the "limited material sub universe" is very great in distance, and impossible to experience at first hand.

    If we were to measure the "great electro-magnetic radiations sub universe, in regards to its "now time", and in regards to its "now motion". We would discover that there is a "universal now time", and a "universal now motion", with respect to the "great electro-magnetic radiations sub universe"—as observed from our vantage point on the Earth.

8   The "great material sub universe", of our entire universe, has "limited time displacements", ie., "variable time displacements". As well as "limited motion displacements; ie., "variable motion displacements". This is the case, in comparison to the "limit and constant time displacements, and limit and constant motion displacements—with respect to the "great electro-magnetic radiations sub universe".

APPENDIX G

THE NEW LAWS OF MOTIVE DISPLACEMENTS

-a-

I would like to state the classical (limited) Newtonian laws of motion in this section—a-, without any comment, or physical interpretations of these three classical laws of motion. And these three laws of motion are called the "limited Newtonian laws of motion". We have:[1]

$$+F_v = 0 \qquad \text{first law of motion}$$
$$+F_v = +\underline{m}\dot{v} \qquad \text{second law of motion}$$
$$+F_v = -F_v \qquad \text{third law of motion}$$

-b-

In this section—b-, I will state the limit classical Newtonian laws of motion, without any comment, or interpretation. And these three laws of motion are called the "limit Newtonian laws of motion". We[2] have:

$$+F_c = 0 \qquad \text{first law of motion}$$
$$+F_c = (+\underline{m}\dot{c} = 0) \qquad \text{second law of motion}$$
$$+F_c = -F_c \qquad \text{third law of motion}$$

-c-

In this section—c-, I will state the limited Einsteinian laws of motion, without comment, nor interpretations on my part. And these six laws of motion are called the "limited Einsteinian six laws of motion". We have:

$+F'_v = 0$ the two first laws of motion

$+'F_v = 0$

$+F'_v = (+\underline{m}\dot{v})/(+1-(\underline{\dot{u}}/\underline{c})^2)^{+\frac{1}{2}}$

                      the two second laws of motion

$+'F_v = (+\underline{m}\dot{v})(+1-(\underline{\dot{u}}/\underline{c})^2)^{+\frac{1}{2}}$

$+F'_v = -F'_v$

                      the two third laws of motion.

$+'F_v = -'F_v$

                          Where $|\dot{v}| = +\underline{\dot{u}}$

-d-

In this section—d-, I will state the limit Einsteinian laws of motion, without comment, nor interpretations, and these six laws of motion are called the: "limit Einsteinian laws of motion.

$+F'_c = 0$

                      the first two laws of motion

$+'F_c = 0$

$+F'_c = ((+\underline{m}\dot{c})/(+1-(\underline{\dot{c}}/\underline{c})^2)^{+\frac{1}{2}} = 0$

                      the two second laws of motion

$+'F_c = ((+\underline{m}\dot{c})(+1-(\underline{\dot{c}}/\underline{c})^2)^{+\frac{1}{2}} = 0$

$+F'_c = -F'_c$

                      the two third laws of motion

$+'F_c = -'F_c$

I have gone beyond the original Newtonian laws of motion in subsection—a-, as well as going beyond the Einsteinian reformation of Isaac Newton's original second law of motion. And instead I formulated our own conception of Newton's first law of motion, his second law of motion, and his third law of motion. I believe these new formulations will be of interest and use to the theoretical scientists, and to applied scientists. Further interpretations of these new laws are given in the reader in the main text of this book.[3]

Notes

[1] With a change of mind, I will give the complete interpretations of these "limited Einsteinian laws of motion". According to my revised Newtonian first law of motion (4), I have given a semantic first law of motion, in which I rejected Newton's first law of motion. This revised first law of motion, that I have given in this "limited Einsteinian laws of motion" is a mathematical formulation, which may not be correct. Since, I had, in this first law of motion, used the mathematic conception of a zero value "0". In which I mathematically considered to represent "inertial rest mass" of a inertial rest mass $+\underline{m}_0$. And also represent "inertial speed displacement".

But this formulation of the first law of motion, mathematically, may not be correct. Hence, we can, as physicists, and applied physicists, can use the semantic revised Newtonian first law of motion, as given in the new law (4).

According to the formulas (11) and (12) in the main text of this book we quote:

"And where $|+\dot{v}| = +\underline{\dot{v}}$. And where the new limited dynamic force measuring formula (11), measures the *instantaneous acceleration displacements* (of increase/of decrease): for a limited non-inertial accelerating displaced mass $+\underline{m}$. And which this same mass $+\underline{m}$ is being "instantaneously impacted", or "continuously impacting" by the dynamic actions of a (classical) limited force: $+F_v$. And these 'dynamic actions' are happening, instantaneously, within a time interval t_b, of one second. Or are happening continuously, within a larger time interval".

"And where the new limited dynamic force measuring formula (12) measures the *instantaneous mass-force displacement* (of increase/of decrease): for a limited non-inertial accelerating displaced mass $+\underline{m}$. And which this same mass $+\underline{m}$ is being "instantaneous impacted", or "continuously impacting" by the dynamic actions of a (classical) limited force: $+F_v$. And these 'dynamic actions' are happening, instantaneously, within a time interval t_b, from (0 to 1), of one second. Or happening continuously, within a larger time interval".

And we also have two more force-measuring formulas, we quote:

"And where $\left|+\dot{v}\right| = +\underline{\dot{v}}$. And where the new limited dynamic force measuring formula (13) measures the *instantaneous volume-force displacement* (of decrease/of increase): for a limited non-inertial accelerating displaced mass $+\underline{m}$. And which this same mass $+\underline{m}$ is being 'instantaneously impacted', or 'continuously impacting' by the dynamic actions of a (classical) limited force: $+F_v$. And these 'dynamic actions' are happening, instantaneously, within a time interval t_b, from (0 to 1), of one second. Or continuously, within a larger time interval".

And where the new limited dynamic force measuring formula (14) measures the *instantaneous density-force displacements* (of increase/of decrease): for a limited non-inertial accelerating displaced mass $+\underline{m}$. And which this same mass $+\underline{m}$ is being 'instantaneously impacted', or 'continuously impacting' by the 'dynamic actions' are happening, instantaneously, within a time interval t_b, from (0 to 1), of one second. Or are happening continuously, within a larger time interval.

[2] In this footnote (2) I will give the "limit Newtonian volume force-law formulas, and will give the limit Newtonian density force-law formulas":

$+F_{\underline{c}V} = 0$

$+F_{cd} = 0$ the first law of volume force formulas, and density force formulas.

$+F_{cV} = +V\dot{c} = 0$

$+F_{cd} = +d\dot{c} = 0$ the second law of volume force formulas, and of density force formulas.

$+F_{cV} = -F_{cV}$

$+F_{cd} = -F_{cd}$ the third law of volume force formulas, and of the law of density force formulas.

I will now give the "limited Newtonian volume force-law formulas and the limit Newtonian density force-law formulas.

$+F_{vV} = 0$

$+F_{vd} = 0$ the first law of volume force law formulas, and the law of density force law formulas.

$+F_{vV} = \underline{V}\dot{v}$

$+F_{vd} = \underline{d}\dot{v}$ the second law of volume force law formulas, and the law of density force law formulas.

$+F_{vV} = -F_{vV}$

$+F_{vd} = -F_{vd}$ the third law of volume force law formulas, and the law of density force law formulas.

[3] I will now give the "limit Einsteinian volume force law formula, and will give the limit Einsteinian density force law formula".

$$+F'_{cV} \quad = \quad 0$$

$$+'F_{cd} \quad = \quad 0$$

The first law of limit volume force law formula, and the first law of limit density force law formula.

$$+F'_{cV} = (+\underline{m}\dot{\underline{c}})/(+1-(\dot{\underline{c}}/\underline{c})^2)^{+\frac{1}{2}} = 0$$
$$+'F_{cd} = (+\underline{m}\dot{\underline{c}})(+1-(\dot{\underline{c}}/\underline{c})^2)^{+\frac{1}{2}} = 0$$

The second law of limit volume force law formula, and the second law of limit density force law formula.

$$+F'_{cV} \quad = \quad -F'_{cV}$$

$$+'F_{cd} \quad = \quad -'F_{cd}$$

The third law of limit volume force law formula, and the third law of limit density force law formula.

I will now give the "limited Einsteinian volume force law formula, and will give the limited Einsteinian density force law formula".

$$+F'_{vV} \quad \geq \quad 0$$

$$+'F_{vd} \quad \geq \quad 0$$

The first law of limited Einsteinian of volume force law formula, and the first law of limited Einsteinian density force law formula.

$$(+\underline{V}\dot{v})/(+1-(\underline{\dot{u}}/\underline{c})^2)^{+\frac{1}{2}} = +F'_{vV}$$

$$(+\underline{d}\dot{v})(+1-(\underline{\dot{u}}/\underline{c})^2)^{+\frac{1}{2}} = +'F_{dv}$$

The second law of limited Einsteinian of volume force law formula, and the second law of limited Einsteinian density force law formula.

$$+F'_{vV} \quad = \quad -F'_{vV}$$

$$+'F_{vd} \quad = \quad -'F_{vd}$$

The third law of limited Einsteinian of volume force law formula, and the third law of limited Einsteinian density force law formula.

The term: V is volume, the term d is density, and the term $\underline{c}$ is speed of light, and term v is limited speed. These formulas, which I have derived from the mass limited formulas. Whereas, $(\underline{d})(\underline{V}) = \underline{m}$.

The first law of limited inertial displacement of a mass, is in regards to an inertial rest mass: $+m_0$, or a a inertial velocity displaced mass. And the symbols greater than zero, or equal to zero, shows this situation.

APPENDIX H

ON THE DYNAMIC LIMIT FORMULAS
$+\underline{mc}$ AND $+\underline{mc}$

Many physicists use these formulas in their theoretical researches, and yet, I believe these formulas to be incorrect, empirically, as is based upon our new kinematic and dynamic considerations. As well as being based upon our new electro-thermo kinematics and dynamics results in the main text of this book.

Since, we have set the two formulas: $+\underline{mc} = 0$, and: $+\underline{mc} = 0$. A substitute for these two limit formulas, are: $+\underline{mv} > 0$, and $+\underline{m\dot{v}} > 0$, and $+\underline{mv} > 0$, and $+\underline{m\dot{v}} > 0$. And which these four new limited formulas, have all been "impacted" by the use of a motive force: $+F$, and the instantaneous discontinuance of this same force. These limited formulas are to be measured in units of 1 kilogram (0 to 3) (10^4) kilometers per one second.

The reason some physicists have set the four limit formulas to be greater than a zero value "0", is because they have considered the light terms: $+c$, and $+\underline{c}$, to be conversion terms, especially in regards to an inertial rest mass term: $+\underline{m}_o$. And yet, any physical or empirical conversion process, involving a mass term: $+\underline{m}$, cannot equal the speed limit $+\underline{c}$ of light. And this empirical fact must always be true, as based upon our new electro-thermo kinematics and dynamic considerations. And thus, the two limit formulas: $+\underline{mc} > 0$, and $+\underline{mc} > 0$, are empirically false.

To see this fact, let us set the two limit formulas, into limit notation:

$$\text{Lim} + \underline{mv} = (+\underline{mc} = 0) \qquad \qquad (a)$$
$$+\underline{v} \rightarrow +\underline{c}$$

And:

$$\text{Lim} + \underline{mv} = (+\underline{mc} = 0) \qquad\qquad\qquad (b)$$
$$+v \rightarrow +c$$

These two limit notations can be interpreted as follows: Instantaneously, within a time interval $\underline{t}_b$, from (0 to $\underline{1}$), of one second. Or within a larger time interval. And whereas, an applied limited motive force: $+F_v$ is used upon an inertial rest mass $+\underline{m}_o$. And at the instantaneous discontinuance of this same applied limited force: $+F_v$—we then have four limited motive displaced formulas: $+\underline{mv} > 0$, and $+\underline{mv} > 0$, and: $+\underline{m\dot{v}} > 0$, and: $+\underline{m\dot{v}} > 0$. But the limit formulas are set as a zero value "0".

And then, the mass terms of these two limit notations (a), and (b), have now, *instantaneously inertially slides* into being limited $+\underline{v}$, and limited $+v$—as uniform inertial speed displacements, and uniform vector velocity displacements.

The mass terms in these two limit notations (a) and (b) are identical (in quantity of mass) with each other. But the zero expression means: that the mass $+\underline{m}$ has "instantaneously imploded/exploded" into very hot and numerous fragments of mass. And also into extreme and intense electro-magnetic radiation effects. And these fragments of very hot mass are being dispersed into the Cosmos.

We also see by our empirical analysis, that all motion, by all cosmic masses, throughout the entire Cosmos, by itself, is not freely given, *and it is not absolutely fundamental*, according to Nature. The various kinds of motion by all cosmic masses, throughout the entire Cosmos, must originally involved a prior kinematic and dynamic actions of applied-motive forces, and the instantaneous discontinuance of these same applied limited motive-forces.[1] As well as the applications of natural forces, and the continuous actions of these natural forces, and without the discontinuance of these natural forces.

In this empirical sense is why we have made use of this revised Isaac Newton's first law of motion. Since, our revised first law of motion does not take for granted, nor as freely given, according to Nature, of all of the many kinds of motive displacements, by all kinds of cosmic masses, within Nature, or within the entire Cosmos.

And now, a proper empirical correction of these two limit formulas[2] $+\underline{mc} = 0$, and $+\underline{mc} = 0$, is to use the four limited formulas: $+\underline{mv} > 0$, $+\underline{mv} > 0$, and: $+\underline{m\dot{v}} > 0$, and: $+\underline{m\dot{v}} > 0$. as they are measured in units of 1 kilogram (0 to 3) (10^4) kilometers per one second.

In this Appendix H, we have been discussing applied forces, such as applied vector forces, and it was not our purpose to discuss natural forces, such as gravitational forces, or electro-magnetic forces. Or of the weak force, or the strong force. All of these natural forces always seem to be presence: without any kinds of instantaneous discontinuances of these same four natural forces.[3]

Notes

[1]   Whereas, when a previous applied force: $+F_v$ had been used upon the inertial mass $+\underline{m}_o$ in order to set it into a uniform acceleration displacement. And when this same force is instantaneously discontinued, the same mass $+\underline{m}$ has now *instantaneously inertially slides* into being a uniform scalar speed displaced mass $+\underline{mv}$, or into being a uniform vector velocity displaced mass $+\underline{m}v$. And these results are happening in free vacuum.

[2]   The reason why these two limit momentum formulas have the zero value "0", ie., $+\underline{mc} = 0$, and: $+\underline{mc} = 0$, is because of the extreme and intense actions of an applied force upon these two masses. And which would result in the same two masses "imploding/exploding" into burning thermo dynamic and kinematic effects that these two masses would experience. There would be nothing left of the mass, only small fragments of mass, or perhaps only left over atoms, etc., and this scenario explains why we have set these two limit momentum formulas with zero values "0". Unfortunately, many physicists may have implicitly thought these two limit momentum formulas are greater than a zero value "0".

Albert Einstein, and many other physicists have thought that Einstein's famous mass-energy formula, ie., $+E = +\underline{mc}^2$ to be greater than a zero values

"0". Which is not the case, it has a zero value "0". This is because the limit momentum formula: $+\underline{mc} = 0$, equals a zero value. And this limit momentum formula can be used to derive Einstein's famous mass-energy formula: $+E = +\underline{mc}^2$.

3   Applied forces upon mass can be discontinued, but it seems that natural forces are never discontinued. Although, as we have see in Appendix E, there may be weak and strong natural gravitational forces working upon a horizontal plane, (weak), and a vertical plane (strong).

APPENDIX I

ON THE RELATIVISTIC INFINITE MASS INCREASE

According to Albert Einstein's special relativity theory, the limiting uniform scalar speed limit $+\underline{c}$ of light, is considered to be a fundamental uniform speed limit; or a fundamental uniform inertial velocity limit for all real cosmic masses, throughout the entire Cosmos, in free vacuum.

And the empirical reason why this uniform speed limit $+\underline{c}$ of light, as it is measured by one second, is a fundamental limit, is because there would result an *infinite mass increase* for any cosmic mass $+\underline{m}$, if this mass would obtain the speed limit $+\underline{c}$ of light.

The proponents of the special relativity theory, have argued in this relativistic manner, as based upon one of their formulas:

$$\operatorname{Lim}(+\underline{m}_o)/(+1-(\underline{u}/\underline{c})^2)^{+\frac{1}{2}} = (+\underline{m} = +\infty) \qquad \text{(a)}$$
$$+\underline{u} \to +\underline{c}$$

And which this relativistic limit notation (a) means is that the term $+\underline{u}$ approaches the uniform speed limit $+\underline{c}$ of light, as it is measured by one second—the mass $+\underline{m}$ would increase to be an *infinite mass increase displacement*.[1]

To see why an "infinite mass increase" displacement cannot happen—we use the uniform limiting factor: $(+1-(\underline{u}/\underline{c})^2)^{+\frac{1}{2}}$. Where we have: $+\underline{u} \to +\underline{c}$, and then: $(+1-(\underline{u}/\underline{c})^2)^{+\frac{1}{2}} = 0$. Hence, by our limit notation, we have:

$$\text{Lim}(+\underline{m}_o)/(+1-(\underline{c}/\underline{c})^2)^{+\frac{1}{2}} = (+\underline{m}_o)(0) = 0 \qquad \text{(b)}$$
$$+\underline{c} \rightarrow +\underline{c}$$

And where this limiting factor: $(+1-(\underline{c}/\underline{c})^2)^{+\frac{1}{2}}$ pans out to

$(+1-(\underline{c}/\underline{c})^2)^{+\frac{1}{2}} = 0$, a dud.

Notes

[1]   The "infinite mass increase" results from the special relativity theory followed by using the limiting factor: $(+1-(\underline{u}/\underline{c})^2)^{-\frac{1}{2}}$ in an incorrect manner is the reason why physicists had thought there could be the fantastic result of: "relativistic infinite mass displacement". The limiting factor: $(+1-(\underline{u}/\underline{c})^2)^{-\frac{1}{2}}$ is of a "decrease/increase" displacement; but this same limiting factor can be given a "increase/decrease" displacement.

Many scientists would write so much non-sense about this "scientific fact" of the infinite mass increase phenomena. And they would write: "that there is not enough energy, or force, within the entire universe, to make a special spacecraft to reach the speed $+\underline{c}$ of light, in free vacuum". The speed of light is considered to be a fundamental constant for all of Nature. But this may not be the case, there may be other speed limits which are greater than the speed limit $+\underline{c}$ of light, in free vacuum. The particle's accelerators may eventually break the light speed barrier. Would there result a huge bang?

APPENDIX J

TO REINSTATE OUR BASIC POSITION

-a-

I would like to reinstate our basic position. Our basic position is simple: we have rejected most if not all of the special relativity theory's basic conceptions. Conceptions, such as: "the special principle of relativity", including conceptions of "relative reference frames", and the "Lorentz-Einsteinian transformation equations". And since, we have already commented upon why we have so rejected these "relativistic conceptions", (see introduction). We then do not need to comment further on these matters.[1]

This new motive-kinematics, and new motive-dynamics, which we have proposed and advanced in this book, involving the uses of "forces". And thus, in this book, is of a true motive-kinematics, and of a true motive-dynamics, since, it makes use of "forces". And this concept of "forces" is of an original applied force which is set upon an inertial cosmic rest mass $+\underline{m}_o$—in order to instantaneously accelerate displace this cosmic rest mass: $+\underline{m}_o$. And which this moving mass $+\underline{m}$ begins to obtain a high velocity, or a speed displacement, in free vacuum. And there will be "kinematic motive effects", and the "dynamic motive effects" that the moving mass $+\underline{m}$ will experience. We then have investigated the important fact of these cumulative thermo dynamic effects that would have come into play for this moving displaced mass $+\underline{m}$.

It is then obvious that we have rejected the special relativity theory's kinematic and dynamic formulas. This is because our new motive-kinematics, and our new motive-dynamics, is a new approach,

146

a new science. Since, we are interested in the physical, or empirical occurrences, which will happen to any inertial cosmic rest mass: $+m_o$. In which this mass has been uniformly accelerated displaced, by the intense actions of a dynamic force. The moving mass $+\underline{m}$ will begin to obtain high accelerations, or high velocity displacements. And the mass $+\underline{m}$ will begin to experience intense and extreme "thermo dynamic and kinematic effects". We now see the fundamental differences between Albert Einstein's special relativity theory, and this new motive-kinematics, and new motive-dynamics theory, which we have advanced in this book, now becomes obvious.

-b-

I would like to examine why Albert Einstein had not made use of the uniform vector velocity limit $+c$—or of the uniform scalar speed limit $+\underline{c}$, of light, to serve as a fundamental vector limit, or as a fundamental uniform scalar limit?[2] I would like to discover why he had used relativistic conceptions, instead of using a vector velocity limit $+c$, or of a speed limit $+\underline{c}$, of light, to serve as a vector velocity limit, or as a speed limit, in his formulation of his special relativity theory?

Around the years 1887-1904, before Albert Einstein had formulated and advanced his special relativity theory, there was much discussions on the Newtonian conceptions of "absolute space and time".[3] And yet, other new concepts were not available to replace this Newtonian concepts of "absolute space and time". Since, the conception of "relativistic" is the formal negation of the conception of "absolutes". And if the conception of "absolutes" were to be rejected, then what new conceptions would take its place? Albert Einstein had an answer, as his plan was to reject the entire conceptions of "absolute space and time". And to replace it with the new conceptions of "relativistic space and time". Which therefore, included the use of the new conceptions of "relativistic inertial velocities", and of "relativistic times", and so on. This was Albert Einstein's plan. And as we know, his special relativity gained wide spread acceptance by most physicists, and by the general public at large.

And yet, to return to our original question: why did not Albert Einstein recognize that his special relativity theory is, in actuality, based upon the uniform vector velocity limit +$\underline{c}$, or based upon the uniform speed limit +$\underline{c}$, of light—as both terms are measured by one second?[4] Whereas, to serve as a fundamental inertial vector velocity limit +c, or as a fundamental scalar speed limit +$\underline{c}$, of light, for: all the inertial, and non-inertial motive displacements, for all material bodies? To answer these questions, we can only conjecture. Since, there was wide spread discussions about relativistic conceptions, and hence, Albert Einstein had naturally used these conceptions. Because the problem would have been for him: what other conceptions could successfully replace both these conceptions of "relativistic" and "absolutes"? Albert Einstein, nor any other physicists, at that time, found none? And only later when the special relativity theory had now been fully accepted by most physicists, is when they implicitly began to recognize that the uniform inertial speed limit +$\underline{c}$ of light, is a fundamental limit: for all the inertial, and non-inertial displacements for all material bodies, throughout the entire Cosmos.[5] However, to implicitly to recognize an important empirical fact is not the same as explicitly to recognize the fundamental importance of this same fact.

And the reason Albert Einstein had not used the uniform vector velocity limit: +c, or use the uniform scalar speed limit +$\underline{c}$, of light—he instead chose to use the outdated relativistic conceptions. Since, these relativistic conceptions were well known at his time. And while, the uniform vector velocity limit: +c, or the uniform scalar speed limit +$\underline{c}$, of light, to serve as a fundamental motive limit, were only implicitly recognized at his time.

And as Albert Einstein had not based his special relativity theory on the vector limit: +c, nor upon the scalar limit: +$\underline{c}$, of light. Whereas, to serve as a fundamental limits for all inertial, and non-inertial motive displacements. And thus, he could not develop a new motive-kinematics, and new motive-dynamics.

However, as Albert Einstein had totally rejected the Newtonian conceptions of "absolute space, motive, time, and position". And including his rejection of "absolute simultaneity time events" (this was so), since, Albert Einstein had rejected the conception of the: "Absolute Ether", which was to be the absolute reference frame. And therefore, he instead chose to follow a cautious road, and to base his kinematics and dynamics upon relativistic conceptions. He had followed this cautious road,[6] because, the conceptions of: "relative inertial reference frames", and their transformation equations were well known at his time.

Yet we are arguing that Albert Einstein had taken the wrong road, instead, he would have been much closer to the truth, if he had attempted to base or found his new kinematics, and his new dynamics, upon the empirical fact of this uniform vector velocity limit +c—or upon this uniform scalar speed limit +$\underline{c}$, of light, as both are measured by one second. But in order for Albert Einstein to have chosen this new way, he would have needed to reject the conceptions of: "relative inertial reference frames", and to reject their various transformations equations. Yet he was not prepared to do this.

And yet, we have done exactly this. And in a like manner, as Albert Einstein had rejected absolute conceptions, we will continue with his plan of rejection, and thus, reject all relativistic conceptions as not having any physical, or empirical basis or status in Nature. And because of our plan of rejection, we then have formulated a new motive-kinematics, and a new motive-dynamics. Which will, successfully deal with the dynamic effects, and the kinematic effects: and including the thermo dynamic and kinematic effects that an inertial, and non-inertial mass +$\underline{m}$ will experience—as acted upon by a limited force: +F_v.

Special relativity theory, as it was formulated by Albert Einstein, had failed to take into account the fundamental fact that applied force must be used upon any inertial rest mass +$\underline{m}_o$. Whereas, to set it into an uniform acceleration displacement, throughout free vacuum. And special relativity theory had also failed to inquire what are the physical effects upon this now uniformly accelerated displaced mass +$\underline{m}$. Such as, what are the "forces of impact", and the "inertial, and non-inertial forces, or effects" that this same mass +$\underline{m}$ will display and experience? We have found that there are real dynamic-motive effects, and real kinematic-motive effects, as well as real thermo kinematic and dynamic effects which will happen to this same moving mass +$\underline{m}$. And this is where the matter rests[7].

-c-

This new motive-kinematics, and this new motive-dynamics, we are advancing in this book,—is meant to replace the classical Newtonian kinematics and dynamics. In regards to all uniform vector motive displacements, and for all uniform scalar motive displacements—in free vacuum space and time. And most important, our new theory, is also meant to replace the special relativity theory, as was advanced by Albert Einstein, for the same reasons.

In fact, this new motive-kinematics, and new motive-dynamics, we are advancing, is a merger between the Newtonian and Einsteinian approaches: since, it preserves much of what is worthy of both theories.[8] Such as, when it deals with the uniform vector motive displacements, (and with the uniform scalar motive displacements), for all limited cosmic masses, in free vacuum. This new motive-kinematics and new motive-dynamics, these new theories, has enlarged upon both the Newtonian kinematics, and dynamics, as well as enlarging upon the Einsteinian kinematics and dynamics.

We have rejected the use of absolute and relativity conceptions, such as "reference frames", (whether they be relative, or absolute). And we have rejected their various transformation equations. Simply, they were not needed by our new theories of motive-kinematics, and motive-dynamics. As well as our ideas and theories of the thermo dynamic and kinematic effect which we have advanced in this book. Since, we have utilized a fundamental natural phenomenon which exists in Nature, to serve as a fundamental limit for limited motive cosmic masses, which is the uniform inertial vector velocity limit: $+c$—or the uniform inertial speed limit $+\underline{c}$, of light, as both of these terms are measured by one second, in free vacuum. And this facts concludes this Appendix J.

Notes

[1]   Absolute and relative concepts were well known around Einstein's time. Henri Poincare, a great physicists, and mathematician, wrote a book in 1904 called:

Science and Hypothesis, in which he wrote about a "relativity principle" to describe relative motion, and relative time. In his book Poincare questioned "absolute conceptions", except for the "aether", which was assumed to permeate all space and time—and which this aether carried all the light waves. Albert Einstein was not aware of Poincare's "principle of relativity", and Einstein in the years from 1900 to 1905 formulated his own "principle of relativity", and thereby Einstein invented his special relativity theory. And Einstein rejected all absolute concepts, including he rejection of the famous Aether.

2   I may be unfair to Albert Einstein, because he formulated two principles for his special relativity theory. And one of his principles he states that the speed of light $+\underline{c}$ was a fundamental constant for the entire universe. And this principle was new to many physicists, who had not studied the work of James Maxwell, who stated that light was a changing electro-magnetic wave phenomena.

3   It is my conjecture that Albert Einstein had with much thought rejected all "absolute concepts", except for the speed of light, and he kept his "relativity concepts". In Albert Einstein's paper: *The Electro-Dynamics of Moving Bodies*, (now called the *Special Theory of Relativity*). He rejected the absolute concepts of absolute space, time, place, and motion. Which these concepts were the kingpin of Isaac Newton's *Mathematical Principles of Natural Philosophy*. Albert Einstein in his paper had thought that his two principles "the principle of relativity" and "the principle of the constant speed of light $+\underline{c}$", to contradict each other. In Albert Einstein's popular book: *Relativity: The Special and General Theory*. Published by: Dover Publications, Mineola, New York, 2001. Albert Einstein resolved the apparent contradiction by using his reasoning on relative distance, relative time, relative simultaneity, as by the use of the Lorentz-Einsteinian transformation equations.

4   Albert Einstein used the scientific conceptions which were available at his time and place. Scientists did not know, we still do not know, if there were natural phenomena of Nature which had a greater speed $+\underline{c}$ of light, as it is measured by one second. It would have been a great leap of imagination if scientists, Einstein included, that they could use the speed $+\underline{c}$ of light, as a fundamental limit for all moving bodies, within the entire universe.

5   The physicist J. Robert Oppenheimer, in a small popular book, I cannot
     remember the title, made the comment that the speed $+\underline{c}$ of light, was a
     "limit".

6   It would have been a very bold step if Albert Einstein had used the speed
     $+\underline{c}$ of light as a fundamental limit for all material moving bodies. And he
     was "cautious" because absolute concepts, and relative concepts, were well
     known at his time and place. Therefore, it is unfair of me to say that he
     "took the cautious road". He used conceptions which were available at his
     time and place.

7   Albert Einstein made use of the Newtonian first law of motion as simply given,
     or for granted. And thus, all inertial states were simply available, without
     inquiring about the forces which had caused these inertial motive states.

8   The new limit and new limited mechanics uses the Newtonian revised first law
     of motion, as well as the second law of motion, and the third law of motion.
     This new mechanics enlarges upon the second law of motion by the use of
     the two limiting factors: $(+1-(\underline{u}/\underline{c})^2)^{+\frac{1}{2}}$ and $(+1-(\underline{u}/\underline{c})^2)^{-\frac{1}{2}}$. In which
     these two limiting factors I borrowed from Lorentz and Einstein' kinematic
     and dynamic formulas. But these two limiting factors may have been invented
     by Michelso, A.A.; and Morely.

APPENDIX K

LIMITING RATE OF CHANGE FACTORS

Introduction

In this new mechanics we are proposing and advancing in this book, we will make use of two limiting factors: $(1-(\underline{u}/\underline{c})^2)^{+\frac{1}{2}}$ and $(1-(\underline{u}/\underline{c})^2)^{-\frac{1}{2}}$. We should note that these two limiting factors have the range of values, in regards to the $+\underline{u}$ term, from: $(0 < k < 3)$. But the $+\underline{u}$ term never being set the value $+\underline{3}$. These two limiting factors have the unit of measurement of: $(0 < k < 3)$ (10^4) kilometers per one second.

-a-

First we will examine the expression: $(x)/(+1(\underline{u}/\underline{c})^2)^{+\frac{1}{2}}$ *for its range of divisional decrease values*. And we have:

$$(+y_o. / +b_o.) = +a \text{ is a decrease value, } (+y_o > +a) \tag{K.0}$$

Where $(+b_o.) = (+1-(\underline{u}/\underline{c})^2)^{+\frac{1}{2}}$

$$(+y_1. /+b_1.) = +c \text{ is a decrease value, as } (+y_1. > +c) \tag{K.1}$$

Where $(+b_1.) = (+1-(\underline{u}/\underline{c})^2)^{+\frac{1}{2}}$

$$(+y_2. / +b_2.) = +d \text{ is a decrease value, as } (+y_2. > + d) \tag{K.2}$$

Where $(+b_2.) = (+1-(\underline{u}/\underline{c})^2)^{+\frac{1}{2}}$

$(+y_3./ +b_3.) = +e$ is a decrease value, as $(+y_2. > +e)$ (K.3)

Where $(+b_3.) = (+1-(\underline{u}/\underline{c})^2)^{+\frac{1}{2}}$

-b-

Next we will examine the expression: $(x)/(1-(\underline{u}/\underline{c})^2)^{\frac{1}{2}}$ *for its range of divisional increase values.* And we have:

$(+y_4.) / +.b_4) = +f$ is an increase value, as $(+y_4. < +f)$ (K.4)

Where $(+.b_4 = (+1-(\underline{u}/\underline{c})^2)^{+\frac{1}{2}}$

$(+y_5. / +.b_5) = +g$ is an increase value, as $(+y_5. < +g)$ (K.5)

Where $+.b_5 = (+1-(\underline{u}/\underline{c})^2)^{+\frac{1}{2}}$

$(+y_6. / +.b_6) = +h$ is an increase value, as $(+y_6. < +h)$ (K.6)

Where $(+.b_6) = (+1-(\underline{u}/\underline{c})^2)^{+\frac{1}{2}}$

$(+y_7. / +.b_7) = +k$ is an increase value, as $(+y_7. < + k)$ (K.7)

Where $(+.b_7) = (+1-(\underline{u}/\underline{c})^2)^{+\frac{1}{2}}$

Hence, the limiting expression $(X)/(1-(\underline{u}/\underline{c})^2)^{\frac{1}{2}}$ *is called the divisional (decrease/increase) factor.*

-c-

And now we will examine the limiting expression: $(y)(1-(\underline{u}/\underline{c})^2)^{\frac{1}{2}}$ *for its multiplicative increase range of values.* And we have:

$(+y_8. (+b_8.) = +i$ is an increase value, as $(+y_8. < +i)$ (K.8)

Where $(+b_8.) = (+1-(\underline{u}/\underline{c})^2)^{+\frac{1}{2}}$

$(+y_9.)(+b_9.) = +j$ is an increase value, as $(+y_9. < +j)$ (K.9)

Where $(+b_9.) = (+1-(\underline{u}/\underline{c})^2)^{+\frac{1}{2}}$

$(+y_{10}.)(+b_{10}.) = +L$ is an increase value, as $(+y_{10}. < +L)$ (K.10)

Where $(+b_{10}.) = (+1-(\underline{u}/\underline{c})^2)^{+\frac{1}{2}}$

$(+y_{11}.)(+b_{11}.) = +m$ is an increase value, as $(+y_{11}. < +m)$ (K.11)

Where $(+b_{11}.) = (+1-(\underline{u}/\underline{c})^2)^{+\frac{1}{2}}$

-d-

And next we will examine the expression $(y)(1-(\underline{u}/\underline{c})^2)^{\frac{1}{2}}$ *for its multiplicative decrease range of values.* And we have:

$(+y_{12}.)(+.b_{12}) = +n$ is a decrease value, as $(+y_{12}. > +n)$ (K.12)

Where $(+.b_{12}) = (+1-(\underline{u}/\underline{c})^2)^{+\frac{1}{2}}$

$(+y_{13}.)(+.b_{13}) = +o$ is a decrease value, as $(+y_{13}. > +o)$ (K.13)

Where $(+.b_{13}) = (+1-(\underline{u}/\underline{c})^2)^{+\frac{1}{2}}$

$(+y_{14}.)(+.b_{14}) = +p$ is a decrease value, as $(+y_{14}. > +p)$ (K.14)

Where $(+.b_{14}) = (+1-(\underline{u}/\underline{c})^2)^{+\frac{1}{2}}$

$(+y_{15}.)(+.b_{15}) = +q$ is a decrease value, as $(+y_{15}. > +q)$ (K.15)

Where $(+.b_{15}) = (+1-(\underline{u}/\underline{c})^2)^{+\frac{1}{2}}$

Hence, the limiting expression $(x)(1-(\underline{u}/\underline{c})^2)^{\frac{1}{2}}$ *is called the multiplicative (increase/decrease) factor.*

-e-

When we examine the *divisional decrease values* in comparison to the *multiplicative increase values*, then these two limiting factors have a physical significance as we have seen in this book. And the same is true for the *divisional increase values*, in comparison to the *multiplicative decrease values*—and where these two other limiting factors also has a physical significance as we have seen in this book. And in this book, we have given a full kinematic and dynamic interpretations of the *divisional decrease/increase range of values* in comparison to *the multiplicative increase/decrease range of values*.

And also where the two limiting factors: $(+1-(\dot{\underline{u}}/\underline{c})^2)^{+\frac{1}{2}}$, and: $(+1-(\dot{\underline{u}}/\underline{c})^2)^{-\frac{1}{2}}$, has the numerical range of values from: $(0 < k < 3)$, but the term: $+\dot{\underline{u}}$, never being set the value $+\underline{3}$. And these two limiting factors have the unit of measurement of $(0 < k < 3)(10^4)$ kilometers per one second.

And this is the first time, as far as we know of it, that the two limiting factors: $(+1-(\underline{u}/\underline{c})^2)^{+\frac{1}{2}}$, and: $(+1-(\underline{u}/\underline{c})^2)^{-\frac{1}{2}}$, have now been given a full analysis in regards to their range of physical values. Whereas, when it comes to the class of: *increase/decrease* range of values for the limiting multiplicative factor. And for the class of: *decrease/increase* range of values for the limiting divisional factor.[1]

Notes

[1]   What we intended to show by these four sections, is that the limiting divisional factor: $(+1-(\underline{u}/\underline{c})^2)^{-\frac{1}{2}}$ has a *decrease/increase displacements*—when the term: $+\underline{u}$, approaches, but never equals the term $+\underline{c}$, of light, in vacuum. And we have shown this by the use of the limiting factor: $(+1-(\underline{u}/\underline{c})^2)^{-\frac{1}{2}}$, in the

sections—a—and—b-. Thus these formulas, or expressions of sections—c-, and—d-, are called the *divisional decrease/increase* formulas. And the same is true for the limiting *multiplicative increase/decrease* formulas. The 'limits' of these two limited factors are the following $(+1-(c/\underline{c})^2)^{+\frac{1}{2}}=0$, and: $(+1-(\underline{c}/\underline{c})^2)^{-\frac{1}{2}}=0)$. For the limiting factor $(+1-(\underline{u}/\underline{c})^2)^{-\frac{1}{2}}$, can be called: *Divisional (decrease/increase) displacements*, And the limiting factor: $(+1-(\underline{u}/\underline{c})^2)^{+\frac{1}{2}}$ can be called: *multiplicative (increase/decrease) displacements*.

APPENDIX L

REINTERPRETATION OF THE MASS-ENERGY FORMULAS: THE SPATIAL-MOTIVE FORMULAS

-a-

Albert Einstein's total mass-energy formula, such as:

$$+\underline{mc}^2 = +E \tag{L.0}$$

Which this famous formula is the dynamic relation between "mass" and "energy". It is a formula of the conversion of mass into energy. This formula has its form because of the fact that it is derived from the total momentum formula: $+\underline{mc}$.

But since, in the text of this book, we have set the limit momentum formula: $+\underline{mc} = 0$. Hence, the zero value "0" means that the mass $+\underline{m}$ has been converted into hot plasmatic fragments of mass, into intense light effects, and burning electro-thermo fragments of mass being dispersed throughout free vacuum space and time, within the Universe.

Hence, we will offer another mass-energy formula:

$$+\underline{mv}^2 = (+\underline{mc}^2 = 0) \tag{L.1}$$

And placing this formula (L.1) into a limit notation, we have:

$$\text{Lim} + \underline{mv}^2 = (+\underline{mc}^2 = 0) \tag{L.2}$$
$$+\underline{v} \rightarrow +\underline{c}$$

158

Instead of using the formula: $(+\underline{mc}^2)$, as the total limit mass-energy conversion formula[1]—we will instead use the limited mass-energy conversion formula: $+\underline{mv}^2$.

Albert Einstein and his proponents of his special relativity theory have made use of the total mass-energy conversion formula: $+\underline{mc}^2 = +E$. Which they considered that $+E > 0$, for all mass-energy that is converted into each other. But this result is a serious mistake on their part. No physical process involving the speed limit $+\underline{c}$ of light, a universal constant, in free vacuum, which involves a mass $+\underline{m}$, can never be greater than a zero value "0". This is because the zero value "0" means that the mass $+\underline{m}$ have been converted into various kinds of energy—by the intense actions of a force: $+F_v$ impacting upon the mass $+\underline{m}$. But there is no limited force: $+F_v$ upon a mass $+\underline{m}$ which will cause the approach to the limit $+\underline{c}$, of light, in free vacuum space and time. Neither is a limit force: $+F_c$ upon a mass $+\underline{m}$ which will cause the mass $+\underline{m}$ to reach, or to approach the speed limit $+\underline{c}$, of light, in free vacuum space and time.

We now have the kinetic mass-energy formula:

$$+\frac{1}{2}(+\underline{mv}^2) = +\frac{1}{2}(\underline{mc}^2 = 0) \tag{L.3}$$

And placing formula (L.3) into a limit notation, we have:

$$\text{Lim}\frac{1}{2}(+\underline{mv}^2) = (\frac{1}{2}(+\underline{mc}^2 = 0) \tag{L.4}$$
$$+\underline{v} \rightarrow +\underline{c}$$

The limited formula on the left hand side of formula (L.3), and limit notation (L.4), we can use as the limited kinetic formula for mass-energy conversions. And the zero value "0". of the right hand side of the limit notation (L.4), means that the mass $+\underline{m}$ has been converted into intense thermo plasmatic fragments of mass-energy, which is dispersed throughout free vacuum space and time.

As there is no motion displacements of a mass, without there being a force used in the first place, then all of these above formulas, the mass $+\underline{m}$ used in these formulas, has had a force used on these various masses. The use of these mass-energy formulas, and of their kinetic mass-energy formulas, are based upon the fundamental "concept of a force used"—which operates upon the mass $+\underline{m}$.

And the reason these formulas have a zero value "0", is that the mass $+\underline{m}$ is converted totally into various kinds of mass-energy. Which is the result of a force, or forces, "impacting", or "interacting" with the mass $+\underline{m}^2$.

We will now use the two limiting factors: $(+1-(\underline{u}/\underline{c})^2)^{+\frac{1}{2}}$, and: $(+1-(\underline{u}/\underline{c})^2)^{-\frac{1}{2}}$ to parameter the limits and formulas from (L.1) to (L.4). And we have:[2]

$$(\frac{1}{2}(+\underline{mv}^2)/(1-(\underline{u}/\underline{c})^2)^{\frac{1}{2}} = \text{spa}_1 \tag{L.5}$$

$$(\frac{1}{2}(+\underline{mv}^2)(1-(\underline{u}/\underline{c})^2)^{\frac{1}{2}} = \text{spa}_2 \tag{L.6}$$

And we have:

$$(+\underline{mv}^2)/(1-(\underline{u}/\underline{c})^2)^{\frac{1}{2}} = \text{spa}_3 \tag{L.7}$$

$$(+\underline{mv}^2)(1-(\underline{u}/\underline{c})^2)^{\frac{1}{2}} = \text{spa}_4 \tag{L.8}$$

And we also have the two limiting factors, as follows: $(1-(\underline{c}/\underline{c})^2)^{-\frac{1}{2}}$ and $(1-(\underline{c}/\underline{c})^2)^{+\frac{1}{2}}$, hence, we have:

$$(\frac{1}{2}(+\underline{mc}^2)/(1-(\underline{c}/\underline{c})^2)^{\frac{1}{2}} = \text{spa}_5 = 0 \tag{L.9}$$

$$(\frac{1}{2}(+\underline{mc}^2)(1-(\underline{c}/\underline{c})^2)^{\frac{1}{2}} = \text{spa}_6 = 0 \tag{L.10}$$

$$+\underline{mc}^2)/(1-(\underline{c}/\underline{c})^2)^{\frac{1}{2}} = spa_7 = 0 \tag{L.11}$$

$$(+\underline{mc}^2)(1-(\underline{c}/\underline{c})^2)^{\frac{1}{2}} = spa_8 = 0 \tag{L.12}$$

the formula (L.5) measures the *kinetic mass-energy-speed* displacement (of decrease/of increase) for a mass $+\underline{m}$, under the actions of a force: $+F_v$. And the formula (L.6) measures kinetic mass-energy displacement (of increase/of decrease), for a mass $+\underline{m}$, under the actions of a force: $+F_v$.

The formula (L.7) measures the limited *mass-energy-_speed_* displacement (of decrease/increase), for a mass $+\underline{m}$, under the actions of a force: $+F_v$.

The formula (L.8) measures the *limited mass-energy* displacement (of increase/of decrease), for a mass $+\underline{m}$, under the actions of a force: $+F_v$.

The formula (L.9) measures the *zero limit kinetic mass-energy-speed* displacement (of zero decrease/of zero increase), for a mass $+\underline{m}$, under the actions of a force: $+F_v$. And formula (L.10) measures the *zero limit kinetic mass-energy* displacement (of zero increase/of zero decrease), for a mass $+\underline{m}$, under the actions of a force: $+F_v$.

And the formula (L.11) measures the zero limit kinetic mass-energy displacement (of zero increase/of zero decrease), for a mass $+\underline{m}$, under the actions of a force: $+F_v$. And the formula (L.12) measures the *zero limit mass-energy* displacement (of zero increase/of zero decrease), for a mass $+\underline{m}$, under the actions of a force: $+F_v$.

-b-

We also have the limiting factors: $(1-(\underline{\dot{u}}/\underline{c})^2)^{-\frac{1}{2}}$, and: $(1-(\underline{\dot{u}}/\underline{c})^2)^{+\frac{1}{2}}$. And then we have the following kinetic energy formulas:

$$(1/2)\underline{m\dot{v}}/(+1-(\underline{\dot{u}}/\underline{c})^2)^{+\frac{1}{2}} = spa_9 \tag{L.13}$$

And:

$$(\frac{1}{2}\underline{m}(\dot{\underline{v}})^2(1-(\dot{\underline{u}}/\underline{c})^2)^{+\frac{1}{2}} = \text{spa}_{10}$$ (L.14)

And we also have the limiting factors: $(1-(\dot{\underline{c}}/\underline{c})^2)^{+\frac{1}{2}}$, and: $(1-(\dot{\underline{c}}/\underline{c})^2)^{-\frac{1}{2}}$. We then have:

$$(\frac{1}{2}\underline{m}(\dot{\underline{c}})^2 / (1-(\dot{\underline{c}}/\underline{c})^2)^{+\frac{1}{2}} = \text{spa}_{11} = 0$$ (L.15)

$$(\frac{1}{2}\underline{m}(\underline{c})^2)(+1-(\dot{\underline{c}}/\underline{c})^2)^{+\frac{1}{2}} = \text{spa}_{12} = 0$$ (L.16)

The formula (L.13) is the first derivative of the terms: $+\underline{v}$, and $+\underline{u}$, where $+\underline{v} = +\underline{u}$. And formula (L.13) measures the *kinetic mass-energy-acceleration displacements* (of decrease/of increase), for a mass $+\underline{m}$, under the actions of a force: $+F_v$. And formula (L.14) is also the first derivative of the terms $+\underline{v}$, and $+\underline{u}$, where $+\underline{v} = +\underline{u}$. And formula (L.14) measures the *kinetic mass-energy displacement* (of increase/of decrease), for a mass $+\underline{m}$, under the actions of a force: $+F_v$. The formula (L.15) is the first derivative of the term $+\underline{c}$. And formula (L.15) measures the *zero limit mass-energy-acceleration displacements* (of zero decrease/of zero increase), for a mass $+\underline{m}$, under the actions of a force: $+F_v$. The formula (L.16) is the first derivative of the term $+\underline{c}$. And formula (L.16) measures the *zero limit mass-energy displacements* (of zero increase/of zero decrease), for a mass $+\underline{m}$, under the actions of a force: $+F_v$.

In this Appendix L, we will not give the volume-mass-energy formulas, and nor give the density-mass-energy formulas.[3] The readers of this book, or this Appendix L, can develop these kinds of formulas on their own. And as an example, about these kinds of formulas, involving volume, and density, please see Appendix[4] D.

Notes

[1]   As we cannot use the limit formulas with the squared term: $+\underline{c}$, ie., $+\underline{c}^2$, multiplied into a mass term $+\underline{m}$, or multiplied into a volume term: $+\underline{V}$, or

multiplied into a density term: $+\underline{d}$. And which always equals a zero value "0". Since, the zero value "0" in these same limit formulas, empirically, means that the mass $+\underline{m}$ has been entirely converted into various kinds of energy, which is dispersed throughout space and time, within the Cosmos. And in this case, we must make use of the limited formulas, involving a limited scalar term: $+\underline{v}$. And whereas, to calculate the limited mass-energy phenomena, is being dispersed throughout free vacuum space and time, within the entire Cosmos.

2   The only mass-energy formulas are the limited mass-energy formulas, which we can use to calculate the same mass-energy of these same formulas. But the limit mass-energy formulas always has a zero value "0". And which this zero value "0" for these limit formulas indicates that the entire mass has been converted into various kinds of mass-energy. Such as, intense mass-energy displacements; intense electro-magnetic radiations effects; intense electro-thermo dynamic effects; and so on.

3   We will offer a few examples of the limited-density mass-energy formulas, and of the limited volume-mass-energy formulas: $(\underline{d}(\underline{v}^2)/(1-(\underline{u}/\underline{c})^2)^{+\frac{1}{2}}$, and $(\underline{d}(\underline{v}^2)(1-(\underline{u}/\underline{c})^2)^{+\frac{1}{2}}$, and: $(\underline{V}(\underline{v}^2)/(1-(\underline{u}/\underline{c})^2)^{+\frac{1}{2}}$, and: $(\underline{V}(\underline{v}^2)(1-(\underline{u}/\underline{c})^2)^{+\frac{1}{2}}$. And so on. And we can interpret these "limited mass-energy" formulas as based upon the "volume term: $+\underline{V}$", and the "density term: $+\underline{d}$", with respect to a mass term: $+\underline{m}$

4   These mass-energy formulas are in actuality three dimensional spatial-motive displacement formulas. This is because these formulas have a three dimensional mass term; a three dimensional volume term; and a three dimensional density term; and also a two dimensional scalar speed term: $+\underline{v}^2$, or a two dimensional vector velocity term: $+v^2$.

APPENDIX M

ON THE SPECIAL RELATIVISTIC FORMULAS

Introduction

Albert Einstein, as based upon his special relativity theory, had advanced various relativistic "dynamic formulas"; and "kinematic formulas"; which dealt with the conceptions of as momentum, force-law, kinetic energy, total mass-energy, time dilation, mass infinite increase, length contraction, doppler effects, spacetimes, quantum mechanics, and for many other conceptions.

-a-

And these various kinds of "relativistic dynamics", and of "relativistic kinematics", of his special relativity theory, can be listed as follows:

$$P = +\underline{m}v = (+\underline{m}_o v) / (+1 - (\underline{u} / \underline{c})^2)^{+\frac{1}{2}} \tag{M.0}$$

(As relativistic momentum)

$$\underline{E}_o = +\underline{m}_o \underline{c}^2 \tag{M.1}$$

(As relativistic rest energy)

$$\underline{T}_c = ((+\underline{m}\underline{c}^2) / (+1 - (\underline{u} / \underline{c})^2)^{+\frac{1}{2}} - 1) \tag{M.2}$$

(As relativistic total kinetic energy)

164

$$E_c = +\underline{mc}^2 = (+\underline{m}_o\underline{c}^2)/(+1-(\underline{u}/\underline{c})^2)^{+\frac{1}{2}}$$ (M.3)

(As relativistic total mass-energy)

$$F_v = (+\underline{m\dot{v}})' \;\; = \;\; (+\underline{m}_o\dot{v})/(+1-(\underline{\dot{u}}/\underline{c})^2)^{+1/2}$$ (M.4)

(As relativistic Newtonian law of force)

$$\underline{t} = (t_o)/(1-(u/\underline{c})^2)^{+\frac{1}{2}}$$ (M.5)

(As relativistic kinematic time dilation towards infinite decrease.

$$L = (L_o)(1-(\underline{u}/\underline{c})^2)^2)^{+\frac{1}{2}}$$ (M.6)

(As relativistic kinematic length contraction

-b-

And in regards to these various kinds of relativistic dynamic and kinematic formulas—and including the conceptions of the special relativity theory—we will list six points. Of why we rejected the special relativity theory as being considered a true "dynamic explanations", or as a true "kinematic explanations". Or even being a true dynamic and kinematic descriptions of the various subjects that it pertains to deal with.

(a) Special relativity theory *simply does take as for given, or for granted* the uniform inertial speed displacements of a mass $+\underline{m}$.

Yet the fact that there is motion implies some kind of force which has set a mass into motion displacements. Hence, special relativity theory fails to take into account that some kind of force must be used to set a mass $+\underline{m}$ into a motive-displacements.[1]

(b) Special relativity theory utilizes conceptions, such as: "reference frames", "transformation equations", and various other conceptions, such as: "relative-symmetry", between "relative inertial observers", with respect to "relative inertial uniform speeds, or velocities", and their "relative inertial times", and so on.

And yet, whereas, these various relativity conceptions, as based upon the special principle of relativity, which we have totally rejected—and they are simply considered as being convenient mathematical methods which have been utilized by physicists in their description, and explanation of Nature.

(c) Special relativity theory utilizes the conception of relative inertial rest mass $+m_o$. And this conception is taken for granted, or as simply given by Nature, and also simply given, or for granted by the proponents of the special relativity theory: without them inquiring how a mass $+m$ had achieved its inertial motive state, or its inertial rest mass state.

But as we have rejected the special principle of relativity, then any conception, as based upon that principle is also rejected. In Appendix N, we will show how an inertial uniform velocity displaced mass $+m$ can be reduced to being an inertial rest mass: $+m_o$.

(d) Special relativity theory fails to take into account that before any inertial rest mass $+m_o$, has achieved a uniform velocity displacement that an external applied force must be used upon this inertial rest mass $+m_o$ to set it into a uniform inertial speed displacement, or to set it into a uniform velocity displacement.

And yet, as we have seen by point (a), all uniform inertial velocity displacements, or inertial speed displacements, are solely taken for granted, or as simply given. We reject this kind of reasoning, since, all dynamic problems must involve the use of a force.

(e) Special relativity theory does not question nor take into account the physical, or dynamic effects, (or of the kinematic effects), of the actions of a force: $+F_v$ upon an inertial rest mass $+\underline{m}_0$.

And yet, as we know, as based upon the physical results of impact or collision between two massive bodies, or by the used of an applied force being set upon an inertial rest mass $+\underline{m}_0$. There are always the physical effects, the electro-thermo dynamic effects that an inertial rest mass $+\underline{m}_0$ will experience, by the intense dynamic actions of a force being set upon this same inertial rest mass $+\underline{m}_0$.

(f) Special relativity theory is in actuality only a relativistic kinematic descriptions of uniform inertial velocity displacements, or speed displacements. Or of the uniform inertial time events, and of the uniform inertial spacetime events. The special theory is not a relativistic dynamical theory of these same events. But the special relativity is only a relativistic kinematic theory. And that is all of what it is.

As we have seen by the points—a-,—b-,—c-, and—d-, including our conception of force—these special relativity theory simply takes for granted and for given, that there are these uniform velocity relative reference frames. And thus, it is easily recognized, that in no way does the concept of "force", and the various "effects"[2], dynamic,", and "thermo-dynamic effects", as caused by this force, does not enter into the conceptions and the results of the special relativity theory. And yet, all dynamic theories must utilize this concept of "force" And all dynamic theories must use the concepts of "dynamic effects", "kinematic effects", and "electro-thermo dynamic and kinematic effects", and so on. This is true if any theory is considered to be called a "dynamic theory". The *General Theory of Relativity* is a dynamic theory, since it uses the concept of "natural forces", and "natural space-time fields". And the general theory is a true dynamic theory also involves relativistic conceptions. But Einstein's special relativity theory is not a dynamic theory, since, it is only a misguided relativistic kinematic theory.

We are rejecting totally, all of these relativistic kinematic and dynamic formulas of Albert Einstein's special relativity theory. Such as, we reject: relativistic momentum,

And instead, we will offer new kinematic and dynamic formulas, as they are based upon the premises and arguments, of this new motive-kinematic theory, and of this new motive-dynamic theory. And which we are proposing and advancing in this book. (See the main text of this book about this matter).[3]

Notes

[1]   Although, it must be said in all fairness to Albert Einstein's special relativity theory, that he had taken all uniform inertial velocity (or speed) displacements—and all inertial relative reference frames—*as given, or as for granted*. As was based upon his understanding and his acceptance of Isaac Newton's first law of motion—as all of his proponents accepted Newton's first law. But I have rejected the Newtonian first law of motion. And I have revised Isaac Newton's first law of motion, I made this revised law more complete, and more concise.

And whereas, most physicists of Albert Einstein time, and even in today's time, had also taken for granted uniform inertial relative velocity, (or speed), displacements *for granted, and freely given*. And these people did not believe it was empirically necessary to inquire into the physical, or empirical, actions of a force: $+F_v$ "impacting" upon an inertial cosmic rest mass $+m_o$. (And the instantaneous discontinuance of this same force: $+F_v$). Simply, they had taken all uniform inertial velocity, (or speed), displacements, greater than zero inertial cosmic rest mass state, for granted, or as simply given, as based upon their naive understanding of Isaac Newton's first law of motion. But we must note that all motion, whether it be uniform inertial motion, or uniform acceleration motion, must have had, originally, a force which was used to set upon any inertial relative reference frame—into uniform velocity (or speed), displacements, free vacuum space and time. Even the natural forces, "acting" upon "all there is in the universe", to have caused motive-displacements "for all is in the universe".

However, by these physicists, including Albert Einstein acceptance of all uniform inertial velocity (or speed) displacements, in free vacuum, *as also as given*. This is the truth as based upon their naive acceptance and their understanding of Isaac Newton's first law of motion.

Hence, neither these physicists, and nor Albert Einstein could not then formulate a true kinematics, and a true dynamics, ie., to make a new theory. That is, on the kinematic and dynamic effects, (including electro-thermo dynamic effects), that an applied force: $+F_v$, would have been used upon an inertial rest mass $+m_o$. The moving mass $+\underline{m}$ would then (imploded/exploded) into its mass being converted into mass-energy displacements. There would be extremely hot burning mass fragments, incredible electro-magnetic radiation phenomena happening to the same mass $+\underline{m}$. And we have the discontinuance of this same force: $+F_v$. And which the fragmented mass $+\underline{m}$, (the previous uniform accelerated displaced whole mass $+\underline{m}$, or what's left of it, as it is into being fragments of mass-energy displacement displays. Like fireworks going on in the night sky. Thus, the mass $+\underline{m}$ is fragmented into hot burning mass-energy, with strong electro-thermo dynamic effects. Unfortunately, I have not been able to give the electro-thermo kinematic and dynamic equations for these kinds of effects. However, it may be that we can use the various kinds of spatial formulas, of mass-energy formulas I have given in the text of this book, and the appendices of this book.

2 In Appendix L, we have set all the limit energy-mass formulas as being equal to the value "0", whenever, this same formulas have a mass term: $+\underline{m}$, and uses another term as being the speed limit $+\underline{c}$, of light, in free vacuum space and time. The zero value "0" signifies that the mass $+\underline{m}$ (that is, the mass-energy in these formulas indicates that a real mass $+\underline{m}$ has been "imploded/exploded" with great heat, hot burning mass, hot plasmatic mass, and intense electro-magnetic radiations effects. And these fantastic effects are happening because of the previous dynamic actions of a very strong force: $+F_v$ "impacting", upon an inertial rest mass $+\underline{m}_o$.

These events are happening, instantaneously, within a time interval $\underline{t}_b$, from (0 to $\underline{1}$), of one second. Or are happening, continuously, within, or at the end of a larger time interval.

[3] We should note that the relativistic mass increase formula: $(+\underline{m}) = (+\underline{m}_o) / (+1 - (\underline{u}/\underline{c})^2)^{+\frac{1}{2}}$, where $(+1 - (\underline{u}/\underline{c})^2)^{+\frac{1}{2}} \to 0$, then $+\underline{m} = 0><0$. Which is non-sense. Since the factor $(+1 - (\underline{u}/\underline{c})^2)^{+\frac{1}{2}}$ was implicitly recognized by the many proponents of the special relativity theory to be a *force-factor*. But we reject this formula of Albert Einstein, and we can replace it by the new formula: $(+\underline{m}/\dot{v}) / (+1 - (\underline{\dot{u}}/\underline{c})^2)^{+\frac{1}{2}} = +F'_v$. As well as by the formula: $(+\underline{m}\dot{v})(+1 - (\underline{\dot{u}}/\underline{c})^2)^{+\frac{1}{2}} = +'F_v$. And the limiting acceleration formula: $(+1 - (\underline{\dot{u}}/\underline{c})^2)^{+\frac{1}{2}}$ is a *limiting-factor*. And these results also applies to the volume-force displacements, mass displacements, and density-force displacements, as well as to motion-displacements, in the text of this book.

We should also note the formula force: $(+\underline{m}_o\dot{v}) / (+1 - (\underline{\dot{u}}/\underline{c})^2)^{+\frac{1}{2}} = +F_v$, where $(+\underline{m}\dot{v} \to 0><0)$ (infinite force displacements) and where $(+1 - (\underline{\dot{u}}/\underline{c})^2)^{+\frac{1}{2}} \to 0$. Which again is non-sense.

We should also note the relativistic force-law formula: $(+\underline{m}_o\dot{v}) / (+1 - (\underline{\dot{u}}/\underline{c})^2)^{+\frac{1}{2}} = +F_v$ has two motion factors, one is the by use of the motion term: $+\dot{v}$, and the second is by the use of the factor-force formula: $(+1 - (\underline{\dot{u}}/\underline{c})^2)^{+\frac{1}{2}}$. Which again is non-sense.

APPENDIX N

ON LIMIT INERTIAL COSMIC REST MASS $+\underline{m}_0$

-a-

We are rejecting the conception of "absolute inertial cosmic rest mass", noted by the term: $+\underline{m}_0$, since, we have rejected the concept of "absolutes" to precisely to describe the various kinematic and dynamic events in Nature.

Albert Einstein had also rejected the concept of *absolute cosmic rest mass*, since, he had rejected the Newtonian conceptions of *absolute space, time, place, and motion*.

However, in the place of the concept of *absolute inertial cosmic rest mass*, Albert Einstein, or the proponents of his special relativity theory, had substituted a relativistic conception, which although satisfies the two principles of the special relativity theory, it is, nonetheless, a very confusing conception.[1]

The relativistic conception of: *relative inertial cosmic rest mass*, also noted by the term: $+\underline{m}_0$—is defined, according to relativistic conceptions, as being due, or determined by the *relative-symmetry* between two relative inertial reference frames. And which contains their own relative inertial observers. Such as, if a mass $+\underline{m}$, within a relative inertial reference frame (d), then it can be concluded by the other relative inertial reference frame (d)'—having an observer, that this same mass $+\underline{m}$ is at *relative inertial cosmic rest mass*. With respect to the relative inertial reference frame (d).

Or it can be concluded, according to relativistic conceptions, that if another mass $+m'$ is within the relative inertial reference frame (d)'.

And then, the relative inertial observer within their own relative inertial reference frame (d): the observer in frame (d) can conclude that their own mass +m has a uniform inertial velocity, (or speed) displacement. And which its motion is greater than the inertial cosmic rest mass. And the mass +m is (within the relative inertial reference frame (d)'). This is the truth for the relative inertial observers, within their own relative inertial reference frame (d)'. And vice versa for the same mass +m, with respect to the observers of their own reference (inertial) frames (d) and (d)'

Thus, we see that the relativistic conceptions of *relative inertial cosmic rest mass*, depends upon the relativistic conceptions of: *relative inertial reference frames*; *relative inertial observers*; and *relative-symmetry*. And all of these relative conceptions are based directly upon the two special principles of relativity.

However, in our new limited mechanics and limit mechanics, we are proposing to have rejected the special principle of relativity, and rejected all the relativistic conceptions, which goes along with it.

And instead of a absolute conceptions, or relativity conceptions upon the conception of "inertial cosmic rest mass", we will offer what we call the definition of: *Limit +m₀ inertial cosmic rest mass*.

This new definition for limit +m₀ inertial cosmic rest mass—will be derived by using the concept of applied force, and the inertial velocity (or speed) displaced mass +m. We now have:

$$\text{Lim}_{+v \to 0} + \underline{m}v = +\underline{m}_o \tag{N.0}$$

Where this limit notation simply means that as the uniform inertial velocity +v displacement, or a inertial cosmic *mass +m, approaches zero inertial velocity displacement*. And whereas, by the instantaneous dynamic actions of an applied motive-force $+F_v$—is instantaneously, acting upon the inertial cosmic mass +m. (The force acting at the opposite of the direction of the moving mass +m). And then also at the instantaneous discontinuance of this applied limited force: $+F_v$. And whereas, the mass +m, previously having a motion displacement, has now achieved a limit inertial cosmic rest mass $+\underline{m}_0$ state, in free vacuum space and time.

What we have accomplished is to define, or to derive, empirically, the new uniform inertial cosmic rest mass $+\underline{m}_0$ state, as being the *limit* for a inertial moving mass $+\underline{m}^2$. And whose limited $+v$ inertial vector velocity displacement, approaches zero "0" non-displacement, in free vacuum space and time.[3]

Notes

[1]   The inertial of a mass $+\underline{m}$ is its resistance to be uniformly accelerated by a force". Page 767, *mass*, "McGraw-Hill Encyclopedia of Physics, etc.,"

[2]   "All matter possesses two properties, gravitation and inertial. The property of gravitation is that every material body attracts every other material body. The property of inertial is that every material body resists any attempt to change its motion". Quoted from: "McGraw-Hill Encyclopedia of Physics", *mass*, page 767, second edition, Sybil P. Parker, editor, New York, N. Y. 1993.

[3]   According to my empirical ideas on mass, there are two kinds of inertial mass. One (1), that of a force "impacting" upon an inertial rest mass $+\underline{m}_0$—setting it into a uniform inertial speed displacement, in free vacuum. The second (2) kind of inertial mass, is that of a force "impacting" upon the uniform inertial speed displaced mass $+\underline{m}$, in the opposite direction of the mass $+\underline{m}$—thereby, setting the mass $+\underline{m}$ into being a uniform inertial rest mass $+\underline{m}_0$.

APPENDIX O

TRANSFORMATION EQUATIONS AND OTHER ITEMS

-a-

The limited Lorentz-Einsteinians' transformation equations are:

$$x' = (x - \underline{ut}) / 1 - (\underline{u} / \underline{c})^2)^{+\frac{1}{2}}$$

$$x = (x' + \underline{ut}') / (1 - (\underline{u} / \underline{c})^2)^{+\frac{1}{2}}$$

$$t' = (t - \underline{ux} / \underline{c}^2) / (1 - (\underline{u} / \underline{c})^2)^{+\frac{1}{2}}$$

$$t = (t' + \underline{ux}' / \underline{c}^2) / (1 - (\underline{u} / \underline{c})^2)^{+\frac{1}{2}}$$

$$y' = y$$

$$z' = z$$

These above equations are called the "limited Lorentz-Einsteinians' transformation equations". They are called "limited" because we have the "limit Lorentz-Einsteinians' transformation equations", which are:

$$x'_c = ((x_c - \underline{ct}) / (1 - (\underline{c} / \underline{c})^2)^{+\frac{1}{2}} = 0)$$

$$x_c = ((x'_c + \underline{ct}) / (1 - (\underline{c} / \underline{c})^2)^{+\frac{1}{2}} = 0)$$

174

$$t'_c = ((t_c - \underline{c}x/\underline{c}^2)/(1-(\underline{c}/\underline{c})^2)^{+\frac{1}{2}} = 0)$$

$$\underline{t}_c = ((t'_c + \underline{c}x'/\underline{c}^2)/(1-(\underline{c}/\underline{c})^2)^{+\frac{1}{2}} = 0)$$

And:

$$y'_c = y_c$$

$$z'_c = z_c$$

And putting all of these equations into limit notation, we have:

$$\underset{\underline{u} \to \underline{c}}{\text{Lim}+} x'_u = (x_c = 0)$$

$$\underset{\underline{u} \to \underline{c}}{\text{Lim}+} x_u = (x_c = 0)$$

$$\underset{\underline{u} \to \underline{c}}{\text{Lim}+} t_u = (t'_c = 0)$$

$$\underset{\underline{u} \to \underline{c}}{\text{Lim}+} t_u = (t_c = 0)$$

These are the "limited equations" Albert Einstein used in his special relativity theory.[1]. And Hermann Minkowski[2], also used these same "limited equations" in his theories on the fourth dimensional spacetime theory. To be more precise, these "limited equations" can be named as being the: "limited *divisional* Lorentz-Einsteinian transformation equations". Our use of the term: "divisional" is because of the limiting factor: $(+1-(\underline{u}/\underline{c})^2)^{-\frac{1}{2}}$ is multiplied into the terms: $(x—\underline{u}t)$; $(x'—\underline{u}t')$; $(t—ux/\underline{c}^2)$; and $(t' + \underline{u}x'/\underline{c}^2)$

And we also have the following new "limited transformation equations", which are:

$$x'_u = (x_u - \underline{ut})(1 - (\underline{u}/\underline{c})^2)^{+\frac{1}{2}} \tag{O.16}$$

$$x_u = (x'_u - \underline{ut}')(1 - (\underline{u}/\underline{c})^2)^{+\frac{1}{2}} \tag{O.17}$$

$$t'_u = (t_u - \underline{ux}/\underline{c}^2)(1 - (\underline{u}/\underline{c})^2)^{+\frac{1}{2}} \tag{O.18}$$

$$t_u = (t'_u - \underline{ux}'/\underline{c}^2)(1 - (\underline{u}/\underline{c})^2)^{+\frac{1}{2}} \tag{O.19}$$

And these new equations are called the "limited multiplicative Lorentz-Einsteinians' transformation equations". And the new "limit multiplicative Lorentz-Einsteinians' transformation equations" are:

$$x'_c = ((x_c - \underline{ct}_c)(1 - (\underline{c}/\underline{c})^2)^{+\frac{1}{2}} = 0) \tag{O.20}$$

$$x_c = ((x'_c + \underline{ct}'_c)(1 - (\underline{c}/\underline{c})^2)^{+\frac{1}{2}} = 0) \tag{O.21}$$

$$t'_c = ((t_c - \underline{cx}_c/\underline{c}^2)(1 - (\underline{c}/\underline{c})^2)^{+\frac{1}{2}} = 0) \tag{O.22}$$

$$t_c = ((t'_c + \underline{cx}'/\underline{c}^2)(1 - (\underline{c}/\underline{c})^2)^{+\frac{1}{2}} = 0) \tag{O.23}$$

And placing these equations into limit notations, we have:

$$\text{Lim} + x'_u = (x'_c = 0) \tag{O.24}$$
$$\underline{u} \rightarrow \underline{c}$$

$$\text{Lim} \, x_u = (x_c = 0) \tag{O.25}$$
$$\underline{u} \rightarrow \underline{c}$$

$$\text{Lim} \, x_u = (x_c = 0) \tag{O.25}$$
$$\underline{u} \rightarrow \underline{c}$$

$$\text{Limt}_{u} = (t_{c} = 0) \qquad\qquad (O.26)$$
$$\underline{u} \to \underline{c}$$

As to what these new limit and limited multiplicative Lorentz-Einsteinians' transformation equations" refers to, I am unable to answer[3]. But perhaps physicists may find them interesting and useful in their researches.

-b-

In the year 1965, a cosmic condition involving low levels of micro-wave radiation was discovered to exist in space and time. And this Cosmic Background Radiation is considered to have originated as a result of the universal event called the Big Bang. However, some respected physicists have proposed, in contradiction to the special relativity theory, that we could measure the true "absolute" uniform inertial velocity displacement of the Earth through this Cosmic Background Radiation. But the proponents of the special relativity theory were only concerned with the global measurement of "absolute" uniform inertial velocity displacement, in which the proponents has denied as a physical possibility. And yet, on the other hand, we are satisfied that a true "discrete absolute" uniform inertial velocity displacement, of the Earth, has now been truly measured for the first time in the Science of Physics.

Although, we do not use the outdated conception of "discrete absolute", but we prefer to use the new concept of "limited v uniform inertial velocity displacement".[4] With respect to the Earth motion through the Cosmic Background Radiation. And these facts should have been another thorn in the side of these proponents of the special relativity theory—yet they choose to ignore it, or to conveniently explain it away.

Because of the confusing results implied from the special relativity theory, such as the conception of "relative-symmetry", we are rejecting entirely as being totally erroneous. And our rejection includes the Lorentz-Einsteinians' transformation equations. And including various other conclusions derived from the premises and arguments of the special relativity theory.

Yet, if we reject the relativistic interpretations of the Lorentz-Einsteinians' transformation equations, then what place do these equations have in the Science of Physics? We are simply putting the Lorentz-Einsteinians' transformation equations where they belong, back into the mathematical methods of the Science of Physics. The Lorentz-Einsteinians' transformation equations, including the concept of relative inertial reference frames, are only convenient mathematical methods. And which may be of some use to account for and to measure various physical events in Nature. And whereas, relative inertial reference frames, do not possess any physical basis in Nature.[5]

Although, the concept of relative inertial reference frames, as a convenient mathematical method, must be reinterpreted as based upon our new limit mechanical theory. Since, we cannot utilize the special relativity theory's conceptions. Our new reinterpretation is as follows: all relative inertial reference frames, which contains their imaginary relative inertial observers, are either limited $+v$ inertial reference frames, or are limited $+\dot{v}$ non-inertial reference frames. And which contains their own limited $+\dot{v}$ non-inertial observers—with respect to the uniform vector velocity limit $+c$—or with respect to the uniform speed limit $+\underline{c}$ of light, in free vacuum.

This means the Lorentz-Einsteinians' transformation equations must also be reinterpreted, as based upon our new limit mechanical theory. And the Lorentz-Einsteinians' transformation equations are simply: "limited $\underline{v}$ transformation equations". Which involves the mathematical concept of limited $\underline{v}$ inertial reference frames. And which contains their own imaginary limited inertial observers.

And as for their coordinate terms: (x) and (x')—and the time coordinate terms (t) and (t'); these terms, as they are related by the limited $\underline{v}$ transformation equations. And they can only refer to the limited $\underline{v}$ order relation of: "zero, equal, less, and greater". With respect to the coordinate terms of (x) and (x')—and also with respect to the time coordinate terms of (t) and (t').

And thus, we see, mathematically, if physicists want to make use of a transformation between two limited $\underline{v}$ inertial reference frames (p) and (p')—and which contains their own imaginary limited $\underline{v}$ observers (e) and (e'). And then to describe and to explain this transformation, the limited $\underline{v}$ order relations of: "zero, equal, less, and greater" must be utilized. And this new reinterpretation is based directly upon the new limit mechanics we are proposing and advancing in this book. But let us hope, that if these newly reinterpreted Lorentz-Einsteinians' transformation equations, were to be utilized by physicists: they then must not make the mistake in believing their conclusions are then empirically true about the facts and events of Nature. Since, as we have seen, Nature herself, does not make use of the concept of reference frames, nor of any mathematical transformation devices. Mathematically, these newly reinterpreted Lorentz-Einsteinians' transformation equations are only a convenient mathematical method. For physicists, in their attempt to derive empirical correlations between the physical phenomena within Nature. And only experimental evidence will either confirm or refute the physicists' mathematically derived conclusions about Nature.

According to our new limit mechanical theory, we will not make use of these outdated mathematical conceptions of: "relative inertial reference frames"; their: "transformation equations"; and "relative inertial observers", and the like. Because these conceptions are not needed in our new limit mechanical theory. And instead, we will rely upon the new concepts of "limit" and "limited". In regards to the uniform motive displacements for all limited cosmic masses. And which is in regards to the constant vector velocity limit +c, or to the speed limit +$\underline{c}$, of light, as it is measured by one second, in free vacuum.

Although, physicists and engineers will continue to use "primed coordinated systems", and their: "limited inertial reference frames", and thus, they will continue to use the "limited transformation equations", and also: "limited inertial observers, or non-inertial observers". They will continue to use these outdated methods and procedures[6], since, most of our current physics are based upon these outdated conceptual, or mathematical methods. Yet we hope that they will reinterpret their primed coordinated system, and their transformation equations. As based upon these new concept of "limit" and "limited", which we have been advancing in this book. However, even then, physicists should not make the mistake of considering that their results are always physically valid and true, when it comes to the correct: and true descriptions of the uniform motive displacement events, throughout the universe.[7]

Since, as we have seen, the special relativity theory, as based upon its various relativistic conceptions, have implied numerous physical events, which were not necessarily, in an empirical sense, supported nor confirmed by the facts of Nature.[8]

And it should be noted that we must not make the mistake of considering that the electro-magnetic radiation phenomena, within Nature, as being some kind of reference frame. Since, all electro-magnetic radiations, having frequencies and wave-lengths. And which we can measure, and which travels throughout space and time, at the uniform speed limit +$\underline{c}$, of light, as it is measured by one second.

In our new limit mechanical theory, we have used the mathematical concept of "scalar" and "vectors". Which usually, can be placed into specific primed coordinate systems. And yet, this: "specific primed coordinate systems"—should not be thought of as having a physical basis in Nature. As it is only a mathematical graphic method, which consists of primed coordinated axes, in order for us to attempt to correlate various uniform inertial velocity (or speed) events within Nature.

-c-

There is one more relativistic result which we are questioning, and it is Herman Minkowski's theories on the 4th dimensional spacetime. We question Minkowski's theories, since, as we have totally rejected the special principle of relativity, in regards to its relativistic conception of: "relative-symmetry" between relativistic time events. And between relativistic inertial velocity motion events within space and time.

As it is well known, Minkowski had used the special principle of relativity, and of its relativistic conceptions of: "relative-symmetry" between relativistic time events—and between relativistic inertial, or non-inertial, motion events within space and time. And he had also made use of the relativistic interpretations of the Lorentz-Einsteinians' transformation equations in his spacetime theories.

Minkowski had discovered, mathematically, that the Lorentz-Einsteinians' transformation equations were unique rotations in a mathematical 4th dimension. However, as we have rejected the special principle of relativity, and the various relativistic conceptions which goes along with it. And thus, as we have reinterpreted the Lorentz-Einsteinians' transformation equations. As based upon this new limit mechanical theory we are proposing. And then, if there is a real physical 4th dimensional spacetime, then the physical, or empirical, conception of it must be developed upon the new ideas, and principles, we have developed in our new mechanical theory. And if this cannot be done, then the Minkowski's 4th dimensional spacetime theory was based upon an imprecise and false theory: the special theory of relativity.[9]

The mathematical conception of coordinated systems, inertial relative reference frames, and transformation equations, are derived directly form Rene' Decartes analytic geometry—which he invented in the year[10] 1628.

But as I have, in this book, rejected the mathematical, and the empirical conception of inertial relative reference frames, as leading to empirical false notions about reality.

I question the empirical conceptions of inertial relative reference frames, the Lorentz-Einsteinians' transformation equations, and of primed coordinated systems. Since, as we have seen, these same conceptions, in the special relativity theory, leads to false results. Instead of using these old fashioned conceptions to measure other inertial relative reference frames—we could use measuring devices such as: radar, laser, maser, sight of eye, and other measuring devices. However, physicists will continue to use primed coordinated terms, and systems, in their equations, but the use of these conceptions are in actuality, very old fashioned ideas and concepts, which can lead to false empirical results.

However, in the year 1908, September 21st, H. Minkowski, a mathematician[11], proposed that time by itself, and space by itself, will forever not be distinct from each other—but will be united into the new mathematical, and empirical conceptions of the: "4th dimensional spacetime".

H. Minkowski developed his theory as based upon the Principle of Relativity, and upon the Lorentz-Einsteinian transformation equations, which Albert Einstein used in his special relativity theory.[12].

H. Minkowski, in his article, proposed a new set of 4th dimensional coordinates for his spacetime theory. These coordinates are:[13]

$$+\underline{c}^2\underline{t}^2 - x^2_{\;c} - y^2_{\;c} - z^2_{\;c} = +\underline{1}$$

According to this new limit mechanical theory, we have proposed in this book, the Minkowski's theory is now called "the limit 4th dimensional spacetime theory". Thus, we now propose a "limited 4th dimensional spacetime theory". These coordinates are:

$$+\underline{v}^2\underline{t}^2 - x^2_{\;v} - y^2_{\;v} - z^2_{\;v} \leq +\underline{1} \qquad\qquad (O.28)$$

Where the term: $+\underline{v}$ is uniform speed displacement, for any limited mass $+\underline{m}$. The coordinated equation (O.27) refers to the limit universe taken as a whole. In regards to its "limit 4th dimensional spacetime continuum". And where the coordinated equation (O.28) refers to the limited internal spacetime of any individual mass taken internally. While the coordinated equation (O.27) refers to all the external mass of the universe taken as a whole. And the "limited 4th dimensional spacetime non-continuum" is of limited internal spacetime of any individual mass, or any individual energy, taken separately. That is, the 4th dimensional spacetime *within* any limited and individual cosmic mass, of the microscopic elements within this same mass creates the "limited spacetime" gravitational force-fields, which is different than the "limit spacetime gravitational force-fields" for the entire universe taken as a whole. And these "limited spacetimes within any individual mass" are the microscopic elements such as "atoms, electrons", "moleculars" within any mass $+\underline{m}$. The previous coordinated equation (O.27) of limit spacetime deals with the entire universe taken as a whole. And it does not deal with the microscopic elements, in regards to its "limited spacetime" within any specific cosmic mass. What we are proposing is that there are two distinct spacetime geometries: in regards to the limited aspects, which is the "limited spacetime geometry" within any cosmic mass, taken individually. And "limit spacetime geometry" of the entire cosmos, in regards to all the mass and energy, of the entire universe.

What we have accomplished with this new model we are proposing is that we have united the two conceptions of "macroscopic" and "microscopic" with each other. This is because in regards to our use of the conceptions of "two distinct and different 4th dimensional spacetimes".

We have arrived at this new model as based upon the two coordinated equations (O.27) and (O.28). These two equations are similar with each other, as they both use the three dimensional terms: $(-x_c—y_c—z_c)$ and $(-x_v—y_v—z_v)$, but these terms are not equal with each other. We also have: $+\underline{ct} > +\underline{vt}$, where placing these two terms into a limit notation, we have:

$$\text{Lim} + \underline{vt} = +\underline{ct} \tag{O.29}$$
$$\underline{v} \to \underline{c}$$

And placing these two and three dimensional coordinated equations with each other into a limit notations, we have:

$$\text{Lim}(-x_v—y_v—z_v) = (-x_c—y_c—z_c) \tag{O.30}$$
$$v \to \underline{c}$$

As based upon these results, we can place the two coordinated equations (O.27) and (O.28), leaving out the terms: $+\underline{ct}$, and $+\underline{vt}$, we then have: $(-x_v—y_v—z_v) > (-x_c—y_c—z_c)$.

According to the limit coordinated equation (O.27), the universal "spacetime gravitation action of limit mass and energy travels at the speed distance of $+\underline{ct}$, in spacetime". But according to the limited coordinated equation (O.28), the "spacetime gravitational action of limited mass and energy travels at the speed distance of $+\underline{vt}$, in limited spacetime". Although, in the limit notation (O.29), we have treated both the terms: $+\underline{vt}$ and $+\underline{ct}$, in regards to their time terms: $+\underline{t}$, as identical or equal, which may not be the case. Hence, if the time $+\underline{t}_v$, of the term: $+\underline{vt}$, is greater than the time $+\underline{t}_c$, of the term: $+\underline{ct}$, ie., $+\underline{t}_v > +\underline{t}_c$. Hence, $+\underline{vt} = +\underline{ct}$, can be a physical possibility. Which this equation means that for all limited mass and energy, the time of gravitational force-field, is longer in time action, and distance action, in comparison to all limit mass and energy, the time of gravitational force-field, is shorter in time action, and distance action.

Notes

[1]   See Albert Einstein's "the principle of relativity", in article: *The Electrodynamics of Moving Bodies*. P. 35-65. Dover Publications, Inc., 31 East 2nd street, Mineola, N. Y.., 1952.

[2]   See Herman Minkowski's *Space and Time*, Ibid, P. 37-91.

[3]   The only use of coordinated systems by physicists is to measure the motion of a mass—is with analytic geometry methods. But in real life our experiences of the geometrical motion of a mass, we can use radar, radio, laser, maser, sight of eye, etc., to determine the speed, acceleration, and perhaps determine the quantity of mass, density, volume, of this same mass.

[4]   Albert Einstein's special relativity theory, in regards to motion, of mass, or masses, has led to empirical invalid results.

[5]   An example is the so-called twin paradox, which physicists were unable to resolve, nor explain, by the principles of the special relativity theory. The twin paradox was, in actuality, an anomaly of the special relativity theory. And this anomaly indicated some thing was wrong or false with the special relativity theory.

[6]   We can extend H. Minkowski's spacetime theory for it to be a more precise and a more correct theory of spacetime.

[7]   See Carl B. Boyer, *A History of Mathematics*, p. 335-346. We should note, that Galileo, may have been the first scientist to have used the idea of "inertial reference frames".

[8]   See again footnote—2-.

[9]   Ibid, see Herman Minkowski, p. 77.

[10] Instead of using the equation (O.28) for the microscopic elements of internal mass, in regards to its gravitational spacetime, we may use the equation:

$$+\underline{c}^2\underline{t}^2 - s^2_c - r^2_c - r^2_c - w^2_c < +\underline{1} \tag{10.0}$$

For the coordinated system of internal microscopic elements of a mass +$\underline{m}$, or for all internal microscopic elements of masses, within the Cosmos. And we now have:

$$-s^2_c - r^2_c - w^2_c < x^2_v - y^2_v - z^2_v \tag{10.1}$$

And the equation (10.0) tells us that the gravitational spacetime of internal microscopic elements of mass, that its gravitational spacetime field travels at the speed limit +$\underline{c}$ of light, in free vacuum. And these gravitational spacetime fields of internal microscopic mass, in regards to its microscopic elements *helps the internal mass to connect together.* See footnote—11-, for more about this thesis.

[11] Our conception of: "internal gravitational spacetime fields", within any bulk mass, is, in actuality, the quantum world, or the quantum sub universe.

In the quantum world, there are elementary particles of Nature. Some of these particles are the moleculars, atoms, nucleus, electrons, protons, neutrons, and many other particles. All of these elementary particles make up what is called the "internal quantum gravitational spacetime fields".

We must note the following principle: "I. whenever there is mass, there is always "internal microscopic gravitational spacetime fields", of any bulk mass, with respect to the quantum world, or with respect to the quantum sub universe. And these "internal spacetime fields" are of the "fourth dimensional spacetime fields". It is my conjecture, that the quantum gravitational spacetime fields—*as caused by the internal quantum particles within bulk mass*—are very much different with bulk mass, taken as it is given in our entire universe. Bulk mass of our universe *has its own external gravitational spacetime fields.* If this is the case, we can ask: "why are there three more forces, the 'electro-magnetic force', the 'weak force', and the 'strong force'"?

11   It is my conjecture that the internal quantum gravitational spacetime force
      fields, within any bulk mass, is very low in gravitational spacetime actions,
      or strengths. The masses of these elementary particles has very low mass.
      Hence, this means that the gravitational spacetime fields is not powerful, or
      strong enough, to keep the electron in its orbit, or to keep the nucleus at the
      center of the atom. Or to keep intact the atoms of the moleculars—in this
      quantum world, or of the quantum sub universe. And thus, these three other
      forces of the quantum world, are used to keep the electron revolving around
      the atom., etc., and to keep the atoms intact with the molecular. The small
      internal quantum gravitational spacetime fields: are only a left over force, a
      remnant, hardly having any influence in keeping the quantum world intact.
      It is only the three quantum forces, 'the electro-magnetic force'. 'the weak
      force', and 'strong force', taken together, which keeps the quantum world, or
      the quantum sub universe intact. See Appendix F, for more of this conception
      of "sub universes".

12   Since Albert Einstein advanced his General Theory of Relativity in the year
      1915—most if not all physicists have considered Albert Einstein's theory
      to apply to the entire universe, and that his theory is a universal theory of
      gravitation. But I have considered that Einstein's theory to only apply to
      bulk matter and bulk energy in regards to our entire bulk universe. And his
      theory does not apply to the internal quantum world, but only to the bulk
      entire universe.

      We should note that the internal quantum world gravitational spacetime
      fields, in regards to the elementary particles is very much different than the
      gravitational spacetime fields of our entire bulk universe.

      We should also note that the internal quantum gravitational spacetime
      force-fields is very weak, and this kind of gravitational spacetime force-fields,
      is only a left over force, a weak remnant gravitating force, for the individual
      particles of the internal quantum sub universe.

      The event called "the big bang" created the elementary particles, and the
      elementary forces, of the quantum sub universe, to order and to control, these
      elementary particles. But I am conjecturing that the gravitational spacetime
      force-fields were hardly a quantum force at all, only a remnant force, with
      respect to these elementary particles, taken individually.

See also, Albert Einstein's *Principle of Relativity*, his article: "The Foundations of the General Theory of Relativity", p. 109-164. Published 1952, in the book by Dover Publications, Inc., 1952, 31st East 2nd street, Mineola, N. Y.

[13] After writing these two above footnotes, I had a thought experiment involving a electron miles above a neutral mass, such as our moon, which does not have a electro-magnetic force-fields. And thus, this thought experiment sent me on wild chases to understand my ideas of quantum gravity, and bulk mass gravity.

To explain this fact of the lone electron being "connected" to the bulk mass of our moon, I can up with two possibilities, which are: (1), that quantum gravity instead of being localized inside of the quantum sub universe, it expands outwards from into the bulk mass sub universe, ie., the "bulk mass", of the stars, planets, galaxies, etc. Hence, all bulk mass gravitational spacetimes force-fields, are only a quantum spacetimes gravitational force-fields. (2). That there are two kinds of gravitational spacetime force-fields, one, the quantum spacetime gravitational force-fields; and two, the Bulk mass universal spacetimes force-fields. About these two points, we have a sub point (2.a), that the quantum spacetimes gravitational force-fields outwards from its quantum sub universe, but with no expansion of the bulk mass universal spacetime gravitational force-field into the entire bulk universe. And to accept this sub-point (2.a), we may be able to explain this fact is similar with the expansion of the electro-magnetic force-fields from the Earth into the entire universe. Hence, all bulk mass spacetime gravitational force-fields are only a quantum spacetime gravitational force-fields phenomenon. But if the other possibility is true: that all Bulk mass spacetimes gravitational force-fields expands, in its own right, from the universe—then we could argue that all elementary particles free from the quantum sub universe, would also be "connected", to any bulk mass, of our entire universe. Hence, this would explain and account for my thought experiment of the lone electron miles above the moon, and being "caught", or "connected" to the moon's gravity.

But, as based upon the coordinated equation: $(\underline{v}^2\underline{t}^2 - x^2_v - y^2_v - z^2_v)$, and $(\underline{c}^2\underline{t}^2 - x^2_c - y^2_c - z^2_c)$, I have conjectured that the speed of interactions of the elementary particles of the quantum sub universe is measured by the first coordinated equation, which is a limited coordinated equation. I also conjectured that the second coordinated equation measures the bulk mass spacetimes universal gravitational force-fields, which is of a limit coordinated equation. Hence, we would then have two kinds of gravitational spacetimes force-fields. The quantum gravity force-field, and the universal balk mass gravity force-fields.

I have conjectured that the quantum sub-universe gravity to be very weak, a left over force, a remnant, force-field. Hence, the bulk mass universal gravitation spacetimes force-fields to be a stronger force that the quantum gravity force-fields, which, as we conjectured, is a very weak force-fields.

Thus, I then chosen point (2), that there are two unique kinds of gravitational phenomena within our quantum sub universe, and our bulk mass sub universe, ie., of our entire bulk mass sub universe. Such as, quantum gravity which remains within the quantum sub universe, and the bulk mass gravity which remains throughout our entire universe. And which this bulk gravity expands outwards from bulk mass at the velocity limit $+\underline{c}$, of light, in free vacuum. And while the quantum gravity remains in the quantum sub universe, and while the interactions within the quantum sub universe happens at a limited speed $\underline{v}$, as is measured by one second, or of a larger time interval.

APPENDIX P

NOTE ON NOMENCLATURE AND UNITS

-a-

we use the scalar term $+\underline{u}$ (with a bar beneath the letter, or in italics) to represent instantaneous uniform speed magnitude displacement). And we use the vector term: $+\mathbf{v}$, (of the letter in bold type to represent instantaneous uniform speed magnitude, and uniform direction displacement). And in general, all letters with a bar beneath the letters are scalar terms, as are scalar terms in italics. And in general, all the letters with a bar above them are vector terms, as are vector terms in bold type.

And the mathematical relation between these scalar and vector terms are:

$$|\mathbf{v}| = \underline{u}, \text{ and } |\mathbf{v}| = \underline{v}, \text{ where } \underline{u} = \underline{v} \tag{a}$$

Which is that the *absolute value* of the vector term: $\mathbf{v}$ (in two vertical bars) is the scalar term $\underline{u}$, or $\underline{v}$, where $\underline{u} = \underline{v}$.

We use a scalar term $\dot{\underline{u}}$ (with a bar beneath the letter, or in italics, and a dot above the letter) to represent the first derivative of the instantaneous uniform speed magnitude displacement, which then, is instantaneous uniform acceleration displacement. And we use a vector term: $\dot{\mathbf{v}}$ (with a bar above the letter, or in bold type, and with a dot above the letter)—to represent the first derivative of instantaneous inertial velocity displacement, which then, is instantaneous acceleration displacements.

And the units of measurements which are used in the text of this book are MSK Units.[1]

In the text of this book, the most basic unit of measurement is the unit for the uniform speed limit c of light, in free vacuum space and time. And this basic unit is: $(3)(10^4)$ kilometers/ per one second. And it is by using this basic unit of measurement that we will "describt, or "paramenter" the conceptions of "momentum" and "force", which are used in the text of this book.

First, we need to give the MKS Units of measurement for momentum, and for force. For momentum, we have: $1 \cdot$ kilogram meters per one second, or we have: $1 \cdot$ kg $\cdot$ m / s. And for force, have: 1 kilogram meters per one second squared, or we have: $1 \cdot$ kg $\cdot$ m / s^2, or $1 \cdot$ kg(kg)^2 / s^2.

And now, the terms and formulas we use in the text of this book, are directly based upon the basic unit of light in free vacuum space and time—which is: $(3 \cdot 10^4)$ kilometers/per one second. And because we use momentum terms and formulas, as well as using form terms and formulas, we must combine (by multiplication), the units of momentum with the units for the uniform speed limit c of light, in free vacuum. And we have:

$$1 \cdot \text{kg·m} / \text{s·}(0 \text{ to } 3)10^4 \cdot \text{km} / \text{s}) = (1 \cdot \text{kg})(0 \text{ to } 3) \cdot 10^4 \cdot (\text{km})^2 / s^2 \qquad \text{(b)}$$

And which we see, is the unit of measurement, not for momentum, but it is the unit of measure for kinetic energy, which is useless for our purposes.[2] And these results tell us that to acquire the unit of measurement for momentum, we must then use the partial unit of measurement for momentum, and we must use the basic unit of measurement of the speed limit c of light, in free vacuum. And we have, when we set the meter term m as equal to the numeral 1, and then set the second term, s, also equal to the numeral 1. And then, we had the unit for momentum: $1 \cdot$ kg m / s, and then we have the partial unit of measurement for momentum, which is: $1 \cdot$ kg.

Hence, we now combine (by multiplication) this partial unit of measurement for momentum, with the unit of measurement for the speed c of light, in free vacuum. And we have:

$1 \cdot$ kg)(0 to 3)$10^4 \cdot$ m / s) = (1 $\cdot$ kg)(0 to 3)$10^4 \cdot$ km / s)   (c)

And which, as we will see, is our new unit of measurement for all the momentum problems in the text of this book.

An also, some of the terms and formulas that we use in the text of this book, consists in using the basic unit of measurement of the speed limit c̱ of light, in free vacuum. Hence, we must combine (by multiplication) the unit of measurement for force with the basic unit of measurement for the speed limit c̱ of light, in free vacuum. And we have:

$(1 \cdot$ kg m / s^2)(0 to 3)($10^4 \cdot$ m / s) = 1 $\cdot$ kg $\cdot$ (0 to 3)(10^4) $\cdot$ m^2 / s^3 (d)

And which again, this unit of measurement is not for force, but it is a unit of measurement for some kind of energy, and which is useless for our purposes. Therefore, again, this result tells us that to acquire the unit of measurement for force, we must use the partial unit of measurement for the speed limit c̱ of light, in free vacuum.—and then combine it with the unit of measurement for force.

And now, we set the meter term m as equal with the numeral 1̱, and the second term s̱ as also equal with the numeral term 1̱—to use with the unit of measurement of speed limit c̱ of light, in free vacuum. Which is: $10^4 \cdot$ km / s.

Hence, now combine (by multiplication) this partial unit of measurement for the speed limit c̱ of light, with the unit of measurement, we have:

$(1 \cdot$ kg $\cdot$ 1 / s)(0 to 3)(km / s) = (1 $\cdot$ kg)(0 to 3)(10^4)km / s^2 (e)

And which is, as we will see, is our new unit of measurement for all the force problems in the text of this book.

We will now give the instantaneous motive displacement terms, in regards to their derivatives, which we will use in this book.

If we have an instantaneous vector displacement term s, which represents instantaneous vector distance. And then, we will now give the instantaneous first derivative of the displacement term: s, and we have:

$$\underset{t \to 0}{\text{Lim}}(d / d\underline{t})s = (\dot{s} = v) \tag{f}$$

And which is the instantaneous uniform vector inertial velocity displacement term: v.

And the second derivative of the instantaneous uniform vector displacement term s, is:

$$\underset{t \to 0}{\text{Lim}}(d / d\underline{t})\dot{s} = (\dot{v}) \tag{g}$$

And which is the instantaneous uniform vector acceleration displacement term: $\dot{v}$.

And the mathematical relation between these scalar and vector terms are as follows:

$$|\dot{v}| = \underline{\dot{u}} = \underline{\dot{v}}, \text{ and, where } \underline{\dot{u}} = \underline{\dot{v}} \tag{h}$$

And where the absolute value of the vector term $\underline{\dot{v}}$ (in two vertical bars) is the scalar terms: $\underline{\dot{u}}$, or $\underline{\dot{v}}$.

Notes

[1]   In this book we use the units of measurements for the motive-kilometers displacements. But in Appendix D, of this book, we do not give the units of measurements of the temporal-kilometers displacements. And nor in the Appendix L, we do not give the units of measurements for the spatial-kilometers displacements—for the mass-energy formulas.

[2]   See the mass-energy formulas in Appendix L, which are the spatial kilometer displacements formulas.

APPENDIX Q

ON THE NATURAL CONSTANTS OF NATURE

-a-

Some of our natural constants of nature uses the speed limit and constant $+\underline{c}$ of light, in free vacuum.

I had the idea of formulating the "limited constants", where this phrase means: "limited variables"—and these limited constants involves the limited speed term: $+\underline{v}$, which is the limited motion of some of the cosmic bodies in free vacuum.[1] Here are some limit constants, involving the term: $+\underline{c}$. We have:

$$h\underline{c} \ / \ Gm_p^{\,2} \tag{Q.0}$$

(which is the gravitational 'fine structure' constant).
And the limited constant of formula (Q.0) is:

$$h\underline{v} \ / \ Gm_p^{\,2} \tag{Q.1}$$

And placing the two formulas (Q.0) and (Q.1) into limit notation, we have:

$$\lim_{v \to \underline{c}} h\underline{v} \ / \ Gm_p^{\,2} = h\underline{c} \ / \ Gm_p^{\,2} \tag{Q.2}$$

And we have the constant:

$$h\underline{c} \ / \ Gm_e^{\,2} \tag{Q.3}$$

$$h / m_p c \qquad \text{(Q.4)}$$

And for the formulas (Q.3) and (Q.4), we have the limited constants:

$$h\underline{v} / Gm_e^2 \qquad \text{(Q.5)}$$

And:

$$h / m_p \underline{v} \qquad \text{(Q.6)}$$

And placing formulas (Q.3) and (Q.4), and the formulas (Q.5) and (Q.6), into limit notations, we have:

$$\lim_{\underline{v} \to \underline{c}} h\underline{v} / Gm_e^2 = h\underline{c} / Gm_e^2 \qquad \text{(Q.7)}$$

And:

$$\lim_{\underline{v} \to \underline{c}} h / m_p \underline{v} = h / m_p \underline{c} \qquad \text{(Q.8)}$$

And we have the limit constant of small times:

$$h / m_p \underline{c}^2 \qquad \text{(Q.9)}$$

And:

$$g_w m_e \underline{c} / h^3 \qquad \text{(Q.10)}$$

And we have the "limited constants" of formulas (Q.9), and (Q.10):

$$h / m_p \underline{v}^2 \qquad \text{(Q.11)}$$

And:

$$g_w m_e \underline{v} / h^3 \tag{Q.12}$$

And placing the formulas (Q.9) and (Q.10), and formulas (Q.11), and (Q.12) into limit notations, we have:

$$\text{Lim } h / m_p \underline{v}^2 = h / m_p \underline{c}^2 \tag{Q.13}$$
$$\underline{v} \to \underline{c}$$

And:

$$\text{Lim } g_w m_e \underline{v} / h^3 = g_w m_e \underline{c} / h^3 \tag{Q.14}$$
$$\underline{v} \to \underline{c}$$

And we have two more limit constants:

$$\underline{1}_p = (Gh / \underline{c}^3)^{\frac{1}{2}} \tag{Q.15}$$

And:

$$\underline{t}_p = (Gh / \underline{c}^5)^{\frac{1}{2}} \tag{Q.16}$$

Formula (Q.15) is of length, and while formula (Q.16) is of time. We now have the "limited constant" of the formulas (Q.15) and (Q.16). We have:

$$\underline{1}_o = (Gh / \underline{v}^3)^{\frac{1}{2}} \tag{Q.17}$$

And:

$$\underline{t}_o = (Gh / \underline{v}^5)^{\frac{1}{2}} \tag{Q.18}$$

And placing the formulas (Q.15) and (Q.16), as well as formulas (Q.17), and (Q.18), into limit notations, we have:

$$\text{Lim } (Gh/\underline{v}^3)^{\frac{1}{2}} = (Gh/\underline{c}^3)^{\frac{1}{2}} \qquad\qquad (Q.19)$$
$$\underline{v} \rightarrow \underline{c}$$

And:

$$\text{Lim } (Gh/\underline{v}^5)^{\frac{1}{2}} = (Gh/\underline{c}^5)^{\frac{1}{2}} \qquad\qquad (Q.20)$$
$$\underline{v} \rightarrow \underline{c}$$

And we have:

$$\underline{F} = \underline{Armc}^2 \qquad\qquad (Q.21)$$

("where $\underline{m}$ is mass of the repelled object, $\underline{r}$ its distance from the repelling body, $\underline{c}$ is the speed of light, then A is a^2 constant with the units m^{-2}")

And we

$$\underline{F}_o = \underline{Armv}^2 \qquad\qquad (Q.22)$$

And placing formulas (Q.21) and (Q.22) into a limit notation, we have:

$$\text{Lim } A\underline{rmv}^2 = \underline{Armc}^2 \qquad\qquad (Q.23)$$
$$\underline{v} \rightarrow \underline{c}$$

And we have:

$$\underline{c}/\underline{w} = \underline{h}/\underline{mc} \qquad\qquad (Q.24)$$

(which is known as Compton's wavelength, where the term: $\underline{w}$ is wavelength). We then have:

$$\underline{v}/w = \underline{h}/\underline{mv} \qquad\qquad (Q.25)$$

(where formula (Q.25) is "limited Compton's wavelength). And placing formulas (Q.24) and (Q.25) into limit notation, we have:

$$\text{Lim } \underline{v} / w = \text{Lim } h / \underline{mv} = \underline{c} / w = h/\underline{mc} \qquad (Q.26)$$
$$\underline{v} \to \underline{c} \qquad v \to \underline{c}$$

-b-

The natural limit constants of nature may not be "made" at an instantaneous moment, or time—and this is what the limit notations seems to suggest. Such as, the limited speed term: $+\underline{v}$ continues to increase, until it reaches the term: $+\underline{c}$, which is the speed limit and constant of light, in free vacuum.

But if the natural limit constants of nature were "made" "instantaneously", then these limit notations may not have and physical, nor empirical validity.[3] I leave it up to physicists to decide upon these conclusions.[4]

Notes

[1] There are other limited motion terms: $\underline{\dot{v}}$, v, $\dot{v}$. As well as the other limit motion terms: $\underline{\dot{c}}$, $\dot{c}$, c.

[2] The quote is from P.C.W. Davies, *The Accidental Universe*, Cambridge University Press, Cambridge, London, 1982. page 11.

[3] The book I used to obtain some of these natural constants is *The Accidental Universe*, P. C. W. Davies, Cambridge University Press, New York, 1983.

[4] I believe it has not been verified as to whether the creation of light, from atoms, and moleculars, is an instantaneous creation of the speed of light, $+\underline{c}$, as measured by one second. And if this is the case then the formulas on the left hand side of these limit notations may not refer to any thing empirically real. But if the creation of light, having the speed limit and constant $+\underline{c}$ beginning from a zero "0" to a value 3, in units of:

(10⁴) kilometers per one second. And thus the formulas on the left hand side of these limit notations has a physical and empirical significance in nature.

This problem is a major empirical problem, which, I believe has not been fully examined by physicists.

APPENDIX R:

ON TWO CERTAINTY PRINCIPLES

Introduction

In the year 1926, Werner Heisenberg, a young German physicist formulated two indeterministic principles, now called the *Uncertainty Principles*:[1] and they are the two mathematical and physical inequalities:

$$\nabla X \nabla P \geq \frac{\hbar}{2} \tag{R.0}$$

$$\nabla E \nabla T \geq \frac{\hbar}{2} \tag{R.1}$$

For the inequality (R.0), Heisenberg argued that we are unable to measure both simultaneously, the position and momentum of a single electron within a atomic structure, say that of a hydrogen atom.

And also, for the inequality (R.1), Heisenberg argued that we are unable to measure both simultaneously, the energy and time (duration) of a single electron within a atomic structure, say that of a hydrogen atom.

Heisenberg argued that the classical measuring device, say that of a microscope, and a light source and its photons, would describe the position of a single electron, but would disturb the momentum of the same electron, in a hydrogen atom. And if we measure the momentum of the same electron, we would disturb the position of the same electron. And it is the same for the energy and time (duration) of a single electron, in a hydrogen atom.

When we attempt to measure simultaneously, the position and momentum of a single electron, in a hydrogen atom, the high frequency of the photon that we use would displace the momentum of the single electron so that we would be able to only measure the position of this single electron. And when we use a lower frequency photon we would displace the position of this single electron, and only measure the momentum of this single electron. In classical physics, on the contrary, we can always measure simultaneously both the position and momentum, and both simultaneously of the energy and time (duration), of any bulk mass in the universe.

The mathematical interpretations of inequalities (R.0) and (R.1) means that the simultaneously measurements of position, momentum, and the simultaneously measurements of energy, time (duration), exceeds the quantum constant of action, noted by the letter:[2] +h, which is Planck's constant, and which uses the mathematical terms ($\geq$). Planck invented this constant +h to explain black body radiation frequencies.

Instead of using the original quantum mechanics to explain the atomic realm, I am proposing a new classical quantum mechanics where it is possible to measure simultaneously both the position and momentum of two identical electrons of two identical hydrogen atoms, consisting of a bulk mass of liquid hydrogen. And the same is the case for the energy and time (duration) of two other identical electrons of two other identical hydrogen atoms, within a bulk mass of liquid hydrogen.

I will now offer two *certainty principles* of the position and momentum of two identical electrons of two identical hydrogen atoms, of a bulk mass of liquid hydrogen. And the same is true for the energy and time (duration) of two other identical electrons, of two identical hydrogen atoms, of a bulk mass of liquid hydrogen. We have:

$$\nabla X + \nabla P \leq \frac{\hbar}{2} \qquad \text{(R.2), and:}$$

$$\nabla E + \nabla T \leq \frac{\hbar}{2} \qquad \text{(R.3)}$$

The inequality (R.2) means that by using two identical electrons of two hydrogen atoms, of a bulk mass of liquid hydrogen, we can measure the position of a single electron of one hydrogen atom by using a low level frequency photon. And by using the same bulk mass of liquid hydrogen, and by using another identical electron, we can then measure simultaneously, the momentum of this other identical electron, by using a high level frequency photon, of another identical hydrogen atom, within the same bulk mass of liquid hydrogen.

And for the inqualily (R.3), we can do the same for the energy and time (duration) of two other identical electrons, within this same bulk mass of liquid hydrogen, ie., in regards to two other indentical hydrogen atoms.

The mathematical and physical interpretations of the certainty principles (R.2) and (R.3), means that the sum of the position and moment of two identical electrons, within two identical hydrogen atoms - can be measured simultaneously, by using two different kinds of photons (of two different frequencies), impacting upon these two identical electrons, within this bulk mass of liquid hydrogen. Or we can measure the two identical electrons at different times.

And for the inequality (R.3), its mathematical and physical interpretations is as follows: that the sum of the energy and time (duration) of two other identical electron within two identical hydrogen atoms - can be measured simultaneously, by again of using two different kinds of photons (of two different frequencies), impacting upon these two identical electrons, within this bulk mass of liquid hydrogen. Or we can measure the two identical electrons at different times.

And the notation terms ($\leq$), means that we can measure the position and momentum of two identical electrons, of two identical hydrogen atoms, in a very high approximate manner, as well as measuring the energy and time (duration) also in a very high approximate manner, that is, less than planck's constant h, which is a quantum constant of action, or of a quantum constant of energy.

The reason why we use bulk mass of hydrogen to simultaneously measure two identical hydrogen atoms, (in regards to their two identical electrons), is according to Heisenberg that we are unable to measure simultaneously, the two states, or four states of a single electron in a hydrogen atom. Classical physicists used bulk mass, or bulk matter in performing their many experiments. Even nuclear physicists use bulk elementary particles in their experiments. And they don't use only one particle at a specific time. The original quantum mechanics, (Bohr's theory), had used the identical atoms of hydrogen to explain its empirical properties. Bohr had in the year 1912, formulated a model of the hydrogen atom, and eventually it was considered incomplete, and the newer original quantum mechanics had replaced it.

Since we can measure simultaneously the position and momentum of two identical electrons, of two identical hydrogen atoms, of a bulk mass of liquid hydrogen atoms - it may be thought that we can measure simultaneously the position and momentum of a single elementary particle. Heisenberg showed that this was an impossibility.

But by using a bulk mass of liquid hydrogen, we can measure simultaneously the position and momentum of two identical electrons of two identical hydrogen atoms, within a bulk mass of liquid hydrogen; ie., by using photon radiations impacting upon these two, or four identical electrons. And since the same is true for the energy and time (duration) of two identical electrons, of two identical hydrogen atoms. And our measurements are empirically valid and true. And we do these measurements by shooting photo radiations at two, or four identical hydrogen atoms, in regards to their two, or four identical electrons.

And since, all the hydrogen atoms are exactly identical within a bulk mass of liquid hydrogen, we then can measure the position and momentum of two equivalent electrons, and of the two equivalent hydrogen atoms, within this bulk mass of liquid hydrogen. And the same is true for our measurement of energy and time (duration) of two other equivalent electrons, for two equivalent hydrogen atoms.

We can now give the mathematical and physical series of the measurements of a physical series, of a bulk mass of identical hydrogen atoms, in regards to their physical series of position and momentum, for these equivalent hydrogen atoms, in regards to their equivalent electrons. We now have: *an extended certainty principle*:

$$\sum_{1}^{n} (\nabla X + \nabla P)_1 + (\nabla X + \nabla P)_2 ... \leq n / 2(\hbar) \qquad \text{(R.4)}$$

And we can now give the mathematical and physical series of the measurement of a physical series, of a bulk mass of identical hydrogen atoms, in regards to their physical series of energy and time (duration) for these equivalent hydrogen atoms, in regards to their equivalent electrons. We now have: *an tended certainty principle*:

$$\sum_{1}^{n} (\nabla E + \nabla T)_1 + (\nabla E + \nabla T)_2 \ldots \leq n / 2(\hbar) \tag{R.5}$$

Since, there are an K (trillion) of hydrogen atoms in a bulk mass of liquid hydrogen, ie., of 27 cubic feet, to measure all of these atoms is an impossibility. But since, all the hydrogen atoms are equivalent with each other, we only need to measure a small amount of liquid hydrogen atoms.

Section (a)

Albert Einstein, and two other physicists, Podolsky, and Rosen, had shown with a thought experiment, called the EPR thought experiment[3], that if two elementary particles had interacted with each other, and then went there separate ways, in opposite directions, within space and time, and within the universe. We would then know the momentum, position, energy, or time (duration), or spin of the first particle, we would also know the momentum, or position, or energy, or time (duration), or spin, of the second particle: however great there distances between these two particles. And these three physicists argued that this was an physical impossibility. Since, this would mean that there is a kind of event called: "action at a distance", which is greater than the speed of light, in vacuum. "Action at a distance" simply happens. But these three physicists concluded that since the original quantum mechanics allowed for this kind of physical event, they then concluded that the original quantum mechanics is incomplete and inconsistent.

But in the 1960s, a physicist, John Bell formulated a theorem he invented, and also invented a set of inequalities, he argued[4] that "action at a distance" is a physical possibility within the universe. Other experimental physicists made a set of experiments to test his theorem, and test his physical inqualities. And they found that his theorem was true, and the experimental theory they had set up was not violated, and thereby, proving the EPR thought experiment was not violated, but the original quantum mechanics was complete and consistent. In short, there was some kind of physical event called: "action at a distance" between two or more particles that had interacted with each other, and went there separate and opposite ways in the universe. These physical events are called: the entanglement theorem, or principle.

I can not prove that the entanglement theorem is physically invalid, and that there is no physical state of entanglement, and that the original quantum mechanics is incomplete and inconsistent. Bell showed that the original quantum mechanics is complete in regards to the entanglement theorem[5]. But the physical conceptions of "action at a distance" strikes me as being a metaphysical conception that has no place in this new classical quantum mechanics that I am proposing. To argue the negation of the entanglement theorem, or principle, we can note that many physical particles are always interacting with each other, and they go their separate ways without our knowing the momentum, position, energy, time, or spin, of the first particle, then influences the exact momentum, position, energy, time, or spin of the second particle. Which happens by a process called "action at a distance" between the first and second particles. But most physicists would argue that Bell's theorem, is empirically valid and true, since, it has been experimentally established as being so. And that the physical inqualities he had set up were violated as being false by these experiments. All I can conclude is that these experiments have been wrongly interpreted, and if we can establish this point, then Bell's theorem would be false, and the physical inqualities he had set up were not violated. But I can not establish this point. I can only argue that the original quantum mechanics is incomplete, and inconsistent, by presenting a negated classical quantum mechanics. "Action at a distance" is the most strange conception that has ever been invented.

Isaac Newton, in his universal gravitation theory had invented the concept of "action at a distance" to explain universal gravitation. Many scientists of his day thought it be be a metaphysical conception aliken to "animalism" of plants, life, or material bodies, and that it was worthless to explain universal gravitation. But Newton won the day, and it was eventually accepted until Albert Einstein can upon the scene with his physical conception of "space-time" of massive bodies upon the fourth dimensional space-time to explain universal gravitation.[6]

Physicists supporting the entanglement prinicple fail to recognize that there are many elementary particles, such as all the free electrons within the universe that if we could measure, or know, that they have the same momentum, position, or energy, and time (duration), or spin, where they are equivalent with each other. This is the truth as they are equivalent electrons, but they had not interacted with each other, or one another, - and since, it is always a possibility to know the equivalent momentum, position, energy, time (duration), or spin, of equivalent electrons within the universe. This is the case since some free electrons within the universe can have the same momentum, position, energy, time (duration), and spin. And for a electron has a spin of -½spin, or +½spin, this means that they are equivalent with the same spin. What we are attempting to argue is that the momentum, position, energy, time (duration), and spin of these electrons can have the same equivalent physical states. This is the case not because they have interacted with each other, and then "entangled" with each other, but all free electrons, (and electrons in atoms), have been created en mass by the universe with non-equivalent, or equivalent quantum states. If this is not the case, why are all electrons within the universe not entangled with each other? Does this conception of "entanglement" work, or does not work, for all free electrons, and for all electrons in atoms? The concept of "entanglement" is an "anomaly"[7] of the original quantum mechanics And the original quantum mechanics is incomplete and inconsistent. This is my argument. And it is the reason why I have offered a new two certainty principles, to describe and explain the physical quantum states of equivalent, and non-equivalent of free electrons, or free particles, etc., within bulk mass, and within the entire universe. I believe the experiments that had been set up, or constructed to test Bell's theorem, and his inequalities have been wrongly interpreted. Or the experimental devices was incorrectly set up, or constructed.

In the creation of the universe, one electron was not created at one specific time, but all electrons were created en mass; that an extreme number of electrons were created into existence. I consider the entanglement principle to be an anomaly of the original quantum mechanics. And the negation of the entanglement principle to be valid and true. We can formulate the:

First negation of the entanglement Principle:

All the electrons within the universe, can be equivalent with each other, or non-equivalent with each other, in regards to their momentum, position, energy, time (duration), and spin. Even though they had not interacted with each other, except at the beginning of the universe, when they were created en mass by the early universe.

And the experiments showing entanglement as a fact, I believe these experiments are physically invalid, and they have been incorrectly set up, or constructed, and they are invalid physical results. It may be argued that for all the electrons in the universe, when they came into existence, in the early universe, there were in an entangled state - since, they were created en mass with each other, and had interacted with each other. And it may be argued that since they were entangled with each other, they still remain entangled with each other in the expanding universe. Such a possibility is so far fleched in that why do we need to know the momentum, position, energy, time (duration), and spin of a second particle, as based the first particle, when we know the particle quantum states? This is the case, since all electrons in the expanding universe do not always have the equivalent quantum states. Such as, some particles may have $+\frac{1}{2}$spin state, and other particles may have $-\frac{1}{2}$spin state. Or different momentum states, as well as different position, energy, time, quantum states. If all the electrons were entangled with each other, within the universe, they would have the equivalent quantum states. But this is not the case. Hence, this conception of "entangled quantum states" fails at this point, or at this event.

Physicists supporting Bell's theorem, and showing his inequalities to be violated, are hitching their beliefs by bring into their arguments the strange conception of "action at a distance" to explain the most strange conception of "entanglement". But when the universe was created, and the creations of equivalent, and non-equivalent electrons, they had, I believe, interacted with each other, (but this way not be the case), hence, why are all the electrons within the universe not in an "entangled quantum state"? This fact shows the empirical invalidity of the concept of "action at a distance", and the conept of "interacting entanglement", both of these concepts are anomalies of the original quantum mechanics.

The entanglement principle violate Heisenberg uncertainty principles, since, when two particles interact with each other, they interact simultaneously. And this fact violates the two Heisenberg uncertainty principles. Where according to Heisenberg, we can not measure simultaneously the momentum and position, or energy, and time (duration), simultaneously upon one particle. Whereas, when two particles interact it happens simultaneously. And they separate simultaneously, or at different times, in opposite directions. And Bell had concluded this conception to be a part of the original quantum mechanics. And because of the concept of "entanglement", according to Bell he showed that the original quantum mechanics to be complete and consistent. While the original EPR experiments concluded the original quantum mechanics to be incomplete and inconsistent.

And also the concept of "action at a distance" performs, or acts in a simultaneously manner, since, when two particles interact with each other, they then separate in sopposite directions, and then the concept of "action at a distance" comes into play. Hence, the concept of "action at a distance" violates Heisenberg two uncertainty principles. Whereas, the concept of "action at a distance" between two opposite directed particles violates simultaneous criticism of Heisenberg's two uncertainty principles. These arguments implies that the concept of "entanglement" between two opposite directed particles is violated, and it is empirically invalid and false, or neither true nor false. Where the negations of "entanglement", and "action at a distance" is then empirically valid and true.

The second negated principle:

There is no "entangled state" between two interacting particles, (although there is the fact of interacting particles), and there is no "action at a distance" between two opposite directed particles. And when we measure and know that of the particles of the universe, there are some particles, in vacuum, that have the equivalent physical quantum states. And there are other particles, in the universe, that have their own equivalent quantum physical states - in contrast to the first opposite particles.

This negated principles is empirically valid and true, since, when the universe created the particles, some of the particles were equivalent with each other, in their momentum, position, energy, time (duration), and spin. And there were other particles which were created that had different, yet equivalent quantum states. The entire universe is made up with equivalent physical quantum states. Such as, the equivalent particles of electrons, having their own equivalent quantum states. And protons, having their own equivalent quantum states. And mesons, having their own equivalent quantum states, and so on. Our universe consisted en mass of extremely many equivalent quantum particles which were created when the Big Bang happened.

I will set up a thought experiment which I can show how we are to measure simultaneously, the position and momentum of two equivalent electrons. As well as measuring simultaneously, the energy and time (duration) of two equivalent electrons. We have:

electron source

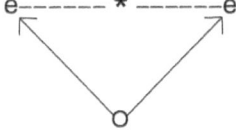

photon source

Two equivalent electrons are shot out from the electron source at simul-times and in opposite directions. And two equivalent photons are also shot out of the photon source at simultaneously times, where they interact with the two equivalent electrons. Therefore, the position and momentum of the two equivalent electrons are measured simultaneously. Or the energy and time (duration) of the two equivalent electrons are measured simultaneously.

As light travels from the past, in space-time continuum of our universe, to the present, or future - then events faster than the speed of light, such as: Simultaneously action at a distance" seems to happen, but this event is false. Since all our measurements done on equivalent particles within the universe only shows that when we measure one equivalent particle as to its quantum states, we then can inductively know the same equivalent quantum states: for the other equivalent particles, which are separated from each other in the space-time continuum of the universe.

When the universe was created, it created an extreme large number Nth of equivalent elementary particles, such as all t he electrons of the total universe, where some of these electrons have the identical or equivalent quantum states, which are separated from each other in this space-time continuum of the total universe.

And if this is the case, then there are no "entangled quantum states" between these identical or equivalent electrons, which are separated in the space-time continuum of the total universe.

Experiments showing the "entanglement principle" to be true have been incorrectly interpreted. And its phenomena of "simultaneously action at a distance" is also incorrectly applied to the total universe, in regards to its equivalent, or identical elementary particles within the space-time continuum of the total universe.

Quoting from Louisa Gilder:[8] "Any time two entities interact, they entangle. It doesn't matter if they are photons (bits of light), atoms (bits of matter), or bigger things made of atoms like dust motes, microscopes, cats, or people. The entanglement persists no matter how far these entities separate, *as long as they don't subsequently interact with anything else* - an almost impossibly tall order for a cat or a person, which is why we don't notice the effect".

"But the motions of subatomic particles are dominated by entanglement. It starts when they interact; in doing so, they lose their separate existence. No matter how far they move apart, if one is tweaked, measured, observed, the other seems to instantly respond, even if the whole world now lies between them. And no one knows how".

There can be interactions with two equivalent particles (or equivalent photons, which we use to measure their entanglement). These influences used upon these two equivalent particles can be due to the electro-weak force, or gravitational force. And even elementary particles, or photons interacting with these two equivalent particles. There is always different kinds of interactions upon these two equivalent particles within the universe. In this sense, there is always different kinds of interactions upon these two equivalent particles, when we attempt to measure their entanglement. This means that the entanglement principle fails for all interacting equivalent particles. Since, there are other influences interacting with these two equivalent particles and thus, the entanglement principles fails for the entanglement experiments.

Even though the large scale gravitation influences may be very weak upon any two equivalent sub-atomic particles, but yet (as yet unknown), there may be the physical process of quantum gravity - and which could influence these two interacting sub-atomic particles, in our testing of their entanglement. It may be possible to isolate the electro-weak force upon these two sub-atomic particles. But can we be sure there are no other influences which would interact with these two entangled sub-particles? I would rather believe in quantum-gravity than that of the principle of entanglement.

Bell had shown that the original quantum mechanics is complete. But I have shown that it is not complete when I advanced my two certainty principles. Heisenberg's two uncertainty princples are invalid and false: when it comes our measurements of two equivalent electrons, in regards to their equivalent quantum states of momentum, and position, of energy and time (duration). Hence, because of our rejection of the original quantum mechanics as being incomplete, with respect to Heisenberg's two uncertainty princples; which we replaced by the two new certainty principles. Hence, we conclude that the original quantum mechanics is incomplete and inconsistent. And the "entanglement principle" fails at this point.

As for the entanglement experiments also fails since, they have been wrongly interpreted and incorrectly explained. These experiments may have been incorrectly set up, or constructed.

Section (b)

In this section (b), I will list some of the paradoxical results of the original quantum mechanics. And while the negations of these paradoxical results are empirically valid and true.

Bohr concept of "complementarity" is a very confusing concept. However, I will use one of the definitions about complementarity as made by Bohr. According to Bohr classical measuring devices creates the quantum world. Without these classical measuring devices, the quantum world not exist. Classical measuring devices complements the quantum world.

Bohr once said, I believe in all seriousness, "there is no quantum world". I believe he meant that without our classical observations of the quantum world not exist. The same is true, according to Bohr, for our measurements of the quantum world.

Bohr concept of complementarity is based directly upon Heisenberg's two uncertainty principles. And according to Bohr, what can not be observed does not exist. And he goes on to say, to observe a quantum phenomena, is to bring it into existence. Some physicists have concluded that our observations of the the universe has created the universe. What Bohr had meant by his statement: "there is no quantum world", is that we cannot measure, nor observe the quantum world by not using our classical measuring devices. Which is false, according to the new certainty principles we have advanced.

Quoting from Jurgen Audretsch, "There is no quantum world and no quantum objects. Only the phenomena are real. Behind it there is nothing. The state vector is purely mathematical auxiliary quantity without correspondence in reality. It serves for the calculation of probabilities of macroscopic events, for example of measurement results . . .The measurement instruments are classical devices, which are not be be described quantum mechanically. The calculations are finally just of providing statements about the classical states of the measuring instruments. The complementarity of position and momentum in the was it is shown by the uncertainty relation has its origin in the fact that no measurement instruments exists for the combined of position and momentum".[9]

To answer this statement we can note that the universal bulk en mass of the entire universe came into existence from the en bulk creations of these elementary quantum particles and their forces. And as we have shown in the introduction of this Appendix R, the complementarity of position and momentum fails for for the measurement of the simultaneously position and momentum. Since, we can by using two identical electrons, we can measure the position of one of the electrons; and then measure the momentum of the other identical electron. And we can do this simultaneously. As for the classical measuring dvices we have used the certainty principles that we have formulated to apply to our classical measurements of the quantum world. Heisenberg's uncertainty principles, or relations fails in our classical measurements of the quanta world. The use of the uncertainty relations is a fundamental anomaly of the original quantum mechanics. The quantum world exists. And it can be measured in higher and higher approximate ways, by the use of newer classical measuring instruments. Physicists have been using one atom at a time to measure other kinds of quantum states. They can even construct a atom with a photon. Which was an impossibility in the early original quantum mechanics era. We live constantly en mass of the en mass bulk quantum world. Look around us of the en mass bulk of our entire universe. En mass bulk of the quantum world exists. It can not be other wise, since if the quantum world did not exist, our entire en mass bulk universe would not exist.

The cat paradox of Schrodinger[10] shows that the micro-scopic physical quantum states influences macro-scopic non-quantum states. And this fact shows that the cat paradox is in actuality an anomaly of the original quantum mechanics. Since, there is a probability of the cat being either alive or dead at a 50% percent of the time. And it is concluded that the cat is both alive and dead at the same probability of 100% percent. These results shows that the basic principles of the original quantum mechanics are incomplete and inconsistent. It is a 100% percent that the cat is both alive and dead, is because it is 50% percent that the cat is alive, and 50% percent that the cat is dead.

And quoting from Colin Bruce:[11]

> "most people have heard of quantum tunneling. Suppose that you fire a particle such as a photon or an electron at some kind of wall that it doesn't have enough energy to penetrate. The wall can be an actual physical barrier such as a thin sheet of aluminium, or a more subtle energy barrier, the equivalent of a hill that the particle does not have sufficient energy to roll up and over. Under these circumstances, the rules of quantum mechanics - specifically, the Heisenberg's uncertainty principles - predicts that because the probability wave associated with the particle slopes over beyond the wall, occasionally the particle will appear to tunnel straight through what would otherwise be an impassable obstacle, just by happening to jump from one part of its probability wave to another."

> "A disturbing feature of quantum tunneling is that it appears to happen instantly. We can describe this in words: 'Any time that you choose to measure the particle, you will find that it is one side of the wall or the other. The implication is that at some point, it must have moved across the wall in no time at all!'"

These quantum situations would not happen if we were to accept the two certainty principles that I have advanced. The original quantum mechanics has strange things associated with it, and this so-called tunneling is a paradox, an anomaly of the original quantum mechanics.[12]

Conclusion

At the end of this appendix R, I finally came up with a postulate R, of what I attempted to explain in the text of his appendix R. We now have:

Postulate R:

When we theoretically measure one elementary particle, as to its quantum state, we can then theoretically deduce the potential measurements of en bulk of all other equivalent elementary particles, in regards to their en bulk classical states.

Let us say that we make a potential measurement of one elementary particle, in regards to its quantum momentum state; hence, we then have measured en bulk all of their equivalent elementary momentum particles, of their total classical momentum states.

We should note that the concept of "quantum" deals with discrete quantum states, while the concept "classical deals with all, or many, or all whole classical states.

If we measure one elementary particle for its quantum momentum, ie., $+\underline{m}k$, where: $+k = 0 = 1/n...1/2...1...\underline{c}$, (where the term: $+\underline{c}$ is the speed number of light in vacuum). And thus, by this elementary momentum formula of its quantum momentum, we then have en bulk mass have measured all of its en bulk quantum and classical momentum states.[13]

The motion of one, or more elementary particle can be classified as follows: $\lim_{0 \to \underline{v}} 0 = \lim_{\underline{v} \to \underline{c}} \underline{v} = \lim_{\underline{c} \to \underline{c}} \underline{c}$ (where $+\underline{c}$ is the speed of light in vacuum).

And we set $+\underline{m}\underline{c} = 0$, which means that the mass $+\underline{m}$ by the action of a force has been converted into energy.

In this sense, classical physics has precedence over quantum physics because it deals with the "whole", the "en bulk" classical states, instead of dealing with individual,, or a single quantum state.

Quantum physicists may measure a single elementary particle, but their measurement only applies to a single model, or a single measurement, with respect to all en bulk equivalent classical particles. Our universe was not created at a single particle at one time, but it was created en bulk mass of all the particles within the universe.

Classical physics, in this sense, has priority over quantum physics. Since, classical physics deal *with the whole, or with en bulk classical states*. While quantum mechanics deal with individual quantum states, taken one at a time.

Many quantum physicists have thought that quantum theory, or quantum mechanics can explain, or be the foundation for classical physics.[14] But the opposite is true, quantum mechanics is founded upon classical physics. It is thought that Plank's constant $\underline{h}$, which is a quantum number, to apply to all black body radiation frequenices. This is the point we are trying to make, Plank's quantum constant, in regards to all black body radiation frequencies but it applies, en bulk, to all en bulk black body radiation frequecies. It is thought that classical physics only applies to continueous classical states, but it also applies to, en bulk, to non-continuous states.

If the entanglement principle is true, it is a classical principle, which deals en bulk with large scale mass and energy. The same is true for all the so-called quantum mechanical "states", in the sense, these "states" are based upon en bulk classical mechanical states.

Also Bell's theorem and inequality, deals with classical mechanical states, instead of individual quantum states. The experimental tests of Bell's theorem and inequality is a probability state, a classical probability state. The probability state of Bell's theorem and inequality is close to the limit zero (0).[15]

Notes

[1]   The principles are now called the uncertainty relations. Heisenberg rejected the simultaneously measurement of position, and momentum, as well as energy and time. Heisenberg concluded that what can not be observed did not exist, or couldn't be real. This is why Mr. Omnes, in his book: *Understanding Quantum Mechanics* talks or writes about "observables", P. 22

[2]   Planck's constant $\underline{h}$, p. 1042, *McGraw-Hill Encyclopedia of Physics* Second edition, NY 1993

3   *Understanding Quantum Mechanics*, p. 54, Roland Omnes, Princeton University Press, New Jersey, 1999

4   *Understanding Quantum Mechanics*, p. 226, Ibid

5   *Understanding Quantum Mechanics*, p. 274, Ibid

6   *The Feynman Lectures on Physics*, Richard Feynman, volume I, 7-8, *Gravity and Relativity*, Addison-Wesley Publishing Company, Inc., Reading, massachusetts, 1989

7   Some quantum physicists think it is proper to use paradoxical results to justify the original quantum mechanics. This is not so. Since, the negations of the paradoxical results are empirically valid and are true, while the original paradoxical results are empirically invalid and are neither true nor false.

8   *The Age of Entanglement*, Louisa Gilder, (no page number given), Alfred A Knopf, New York, 2008

9   *Entangled World*, Jurgen Audretsch, editor, p. 32, Wiley-VCH Gmbh & Co, KgaA

10 *Schrodinger's Rabbits*, Colin Bruce, p. 90, Joseph Henry Press, Washington, D.C. 2004

11 For Mr. Omnes explains away the cat paradox by a principle he calls the decoherence principle, which he explains has been verified in nature. But as far as I am concerned, the cat paradox is an anomaly of the original Quantum mechanics

12 Mr. Omnes, all through his book: *Understanding Quantum Mechanics*, he states that quantum theory is the foundations of classical physics. But the opposite is true. Classical physics is the foundation of Quantum Mechanics. This is because quantum mechanics deals with en bulk of all the elementary particles and their forces.

13 We can use the following numbers: $(0...1/n...1/2...\underline{c})$ for position and for momentum, as well as for energy and time

14 See footnote (12)

15 Some where in Mr. Omne's book: *Understanding Quantum Mechanics*, Ibid, has stated that three millions of photons are used in each test of Bell's theorem and inequality. This is the reason why I considered Bell's results close to the probability limit of zero (0). It has a very low probability.

BIBLIOGRAPHY

G. N. Alekseev
Energy and Entropy
Mir Publishers
Moscow, USSR 1968

Arnold B. Arons
Development of Concepts of Physics
Addison-Wesley Company, Inc.,
Reading, Massachusetts 1965

P. W. Atkins
Quantum
Oxford University Pres
Oxford, Great Britain 1981

David Bohm
The Special Theory of Relativity
W. A. Benjamin, Inc.,
New York 1965

W. Bolton
Motion and Force
Butterworths
London, Great Britain 1980

Max Born
Einstein's Theory of Relativity

Dover Publications, Inc.,
New York, 1965

John Boslough
Masters of Time
A William Patrick Book
Addison-Wesley Publishing Company
Reading, Massachusetts 1992

Werher Von Brun, and:
Frederick I. Ordway III
The Rocket's Red Glare
Archor Press/Doubleday
Garden City, New York 1976

Frederick J. Bueche
College Physics 7ed
Schaum's Outline Series
McGraw Hill Book Company

David C. Cassidy
Uncertainty:
The Life and Science
of Werner Heisenberg
W. H. Freeman and Company
New York, 1982

Beryle E. Clotfelter
Reference Systems and Inertial
Iowa State University Press
Ames, Iowa 1970

Ed; by E. U. Condon &
Hugh Odishaw

Handbook of Physics
McGraw Hill Book Company
New York 1967

ED., by G. K. T. Conn &
G. N. Foeule
Essays in Physics, Volume I
Academic Press
London and New York 1970

G. D. Coughlan and J. E. Dodd
The Ideas of Particle Physics
Cambridge University Press
Cambridge, Great Britain 1991

A. d! Abro
The Rise of the New Physics, Volume I
Dover Publications, Inc.,
New York 1951

Ray D' Invenno
Introducing Einstein's Relativity
Claredon Press
Oxford University Press
Oxford, Great Britain 1992

P. C. W. Davies (his article on):
Essays of Physics: *time*
Addison-Wesley Publishing Company
Reading, Massachusetts 1981

P.C. W. Davies, Ed., of:
The New Physics
Cambridge University Press

Cambridge, Great Britain
(Paperback edition) 1992

Richard Feynman
The Feynman Lectures on Physics,
Volume I
Addison-Wesley Publishing Company
Reading, Massachusetts, Inc., 1963

Ed., by Ben R. Finney, and:
Eric M. Jones
Interstellar Migration and the
Human Experience
University of California Press
Berkeley Los Angeles London 1985

J. T. Fraser
Of Time, Passion, and Knowledge
Princeton University Press
Princeton, New Jersey 1990

Louis Friedman
Starsailing:
Solar Sails and Interstellar Travel
John Wiley & Sons, Inc.,
New York 1988

Irvin D. Gluck
Optics: *the Nature and Applications*
of Light
Holt, Rinehart, and Winston, Inc.,
New York 1964

Victor Guillemin

The Story of Quantum Mechanics
Charles Scribners Sons
New York 1968

Edward R. Harrison
Cosmology: *The Science of the Universe*
Cambridge University Press
Cambridge, Great Britain 1981

Werner Heisenberg
Physics and Philosophy
Harper Torchbook
Harper & Row, Publishers
New York 1962

R. Bruce Lindsay, Ed., of:
Energy: *Historical Development*
of the Concept
Dowden, Hutchinson, & Ross
Stroudsburg, Pennsylvania 1975

L. Marder
Time and the Space Traveller
George Allen, & Unin Ltd;
Great Britain 1971

Arthur I Miller
Albert Einstein: *Special Theory of Relativity*
Addison-Wesley Publishing Company, Inc.,
Reading, Massachusetts 1981

NASA Lewis Research Center
Exploring in Aerospace Rocketry
NASA, Washington D.C. 1971

Sir Isaac Newton
Mathematical Principles of Natural Philosophy
Encyclopedia Britannica, Inc.,
Chicago 1952 USA

Abraham Pais
Subtle is the Lord:
The science and Life of
Albert Einstein
Oxford University Press
New York 1982

Ben Patrusky
The Laser
Dodd, Mead and Company
New York 1966

P. J. E. Peebles
Quantum Mechanics
Princeton University Press
Princeton New Jersey 1992

Han Reichenbach
The Direction of Time
University of California Press
Berkeley Los Angeles London 1971

Wolfgang Rindler
Essential Relativity
Van Nostrand Reinhold Company
New York 1969

Milton A. Rothman

Discovering the Natural Laws
DoubledaY & Company, Inc.,
Garden City, New York 1972

Robert H. Romer
Energy: An Introduction of Physics
W.H. Freeman and Company
San Francisco, California 1976

Michel Shallis
On Time
Schuchen Book
New York 1983

Ed., by Morton Schoor, and
Alfred J. Zaehringer
Solid Rocket Technology
John Wiley and Sons, Inc.,
New York 1967

P. S. Smith & R. C. Smith
Mechanics
John Wiley & Sons
New York, N. Y. 1968, 1990

Murray R. Spiegel
Theory and Problems of Theoretical
Mechanics
Schaum Publishing Co.,
New York 1967

Edwin F. Taylor &
John Archibald Wheeler
Spacetime Physics

W.H. Freeman and Company
San Francisco, California 1966

Stephen Toulmin & June Goddfield
The Fabric of the Heavens
Harper & Row, Publishers
New York, 1961, 1965

Steven Weinberg
The First Three Minuts
Basic Books, Inc.,
New York 1977

Ibid.,
Dreams of a Final Theory
Pantheon Books
New York 1992

Ed., by L. Pearce Williams
*Relativity Theory: Its Origins and
Impact on Modern Thought*
John Wiley & Sons, Inc.,
New York 1968

INDEX

www.ingramcontent.com/pod-product-compliance
Lightning Source LLC
Chambersburg PA
CBHW031835170526
45157CB00001B/310